Francisco Italo Gomes Paiva
Francisco de Oliveira

Potassium and calcium fertigation management in tomato cultivation

Francisco Italo Gomes Paiva
Francisco de Oliveira

Potassium and calcium fertigation management in tomato cultivation

Tomato plants grown in a protected environment and subjected to salt stress

ScienciaScripts

Imprint

Cover image: www.ingimage.com

This book is a translation from the original published under ISBN 978-613-9-61875-0.

Publisher:
Sciencia Scripts
is a trademark of
Dodo Books Indian Ocean Ltd. and OmniScriptum S.R.L publishing group

120 High Road, East Finchley, London, N2 9ED, United Kingdom
Str. Armeneasca 28/1, office 1, Chisinau MD-2012, Republic of Moldova, Europe
Printed at: see last page
ISBN: 978-620-7-70672-3

SUMMARY

OFFER

To my parents **Adailton Pinheiro de Paiva** and **Albecina Gomes da Silveira Paiva** for teaching me the true values of life and for showing me the right path to follow in the pursuit of happiness. And for all their love and support during these years of study.

To my **fiancée Priscila Mendes Targino,** who has always been by my side and little by little we have achieved our goals, because together we are stronger. Someone I can count on in life.

BIOGRAPHY

FRANCISCO ITALO GOMES PAIVA was born in the city of Apodi in the state of Rio Grande do Norte on June 7, 1988. He graduated in Agronomic Engineering from the Federal Rural University of Semi-Arid in March 2013. In 2014 he joined the Postgraduate Program in Soil and Water Management, also at the Federal Rural University of Semi-Arid.

ACKNOWLEDGMENTS

To GOD, for another victory and for all the blessings, willpower, protection and determination on this long road of life.

The Federal Rural University of Semi-Arid for graduating me in Agronomy, a degree in which I recognize my true vocation. And to this institution's Postgraduate Program in Soil and Water Management, where I had the opportunity to grow as a human being and as a professional.

To my advisor Francisco de Assis de Oliveira, a professional whom I admire and respect, for all his teaching, attention, patience and trust over the last two years that we have worked together. For the guidance he has given and for all his support in carrying out this work, showing me that titles alone do not build a man up, but rather the sincere and pleasurable commitment to what he was destined to do.

Marcelo Tavares, advisor and friend who took me in from the beginning to the end of my degree, whom I admire greatly.

Lucas Ramos, a friend and fellow researcher, with whom I shared knowledge and good stories as aspiring researchers. A little man with a big heart.

To my friend Andygley Fernandes for all his support and companionship. A sincere friendship that I will always carry with me.

To my friends Raiza Lopes, Ricardo Carlos, Priscila Maia, Rita de Càssia and Lilia Souza for all their help and availability during the experiment.

To the professors Rafael Oliveira, Jeane Portela, Francismar de Medeiros, Carolina Malala, Nildo Dias, Miguel Ferreira Neto, Neyton Miranda, Fabio Henrique for being transmitters of knowledge and for the example that one day we can achieve our goals, all it takes is determination.

I would like to thank all the people who directly or indirectly helped me to complete this work.

Thank you very much!

SUMMARY

FRANCISCO ITALO GOMES PAIVA. **FERTIGATION MANAGEMENT**

POTASSIUM AND CALCIUM IN TOMATO CROPS GROWN IN A PROTECTED ENVIRONMENT AND SUBJECTED TO SALT STRESS. Mossoró- RN. Federal Rural University of Semi-Arid, February 2017. Dissertation. Postgraduate Program in Soil and Water Management. Advisor: Francisco de Assis de Oliveira.

The production and quality of tomato fruit is directly related to the quality of the water used for irrigation and the adequate supply of nutrients, especially potassium and calcium. This work was carried out at the Department of Environmental and Technological Sciences of the Universidade Federal Rural do Semi-Arido (UFERSA), in Mossoró, RN, with the aim of evaluating the cultivation of tomatoes (*Lycopersicumes culentum* Mill) in a protected environment, using irrigation with saline water and fertigation with different potassium:calcium ratios. The experimental design used was randomized blocks, in a 5 x 4 factorial scheme, with four replications. The experiment was carried out in a protected environment using coconut fiber substrate. The treatments were made up of a combination of five ionic ratios of potassium and calcium (F1-K^+/Ca^{2+} =1.5:1; F2-K /Ca^{+2+} =1.25:1; F3-K /Ca^{+2+} =1:1; F4- K /Ca^{+2+} =1:1.25 and F5-K /Ca^{+2+} =1:1.5) using standard solutions for each crop, according to specialized literature, with four levels of irrigation water salinity (S1-0.5; S2-2.0; S3-3.5 and S4-5.0 dS m^{-1}). The ratio K /Ca^{+2+} =1:1 corresponds to the concentration of these nutrients recommended for the crop in hydroponic tomato cultivation. The plants were evaluated for growth variables (height, stem diameter, number of leaves, leaf area, dry mass of stem, leaves, fruit and total), yield (number of fruits per plant, average fruit mass and fruit production per plant) and fruit quality (soluble solids, total acids, °Brix, Vitamin C and AT/SS ratio). The K^+/Ca^{2+} ratios studied did not mitigate the effect on the variables analyzed. The use of low K /Ca^{+2+} ratios increased the effect of salinity on the plants.

Keywords: Solanaceae, use of saline water, mineral nutrition.

1. INTRODUCTION

The tomato plant is grown in almost every state in the country, dividing its production into table tomatoes and tomatoes for processing. However, even with the growth in the area under industrial tomato cultivation, most of the production is destined for *fresh* consumption (MATTEDI et al., 2007).

The tomato production chain is one of the most important in the food industry, with production being one of its initial links. According to a study on the "Development of the Tomato Agro-Industrial System", due to the intrinsic characteristics of production, processing and marketing, tomato crops are destined for fresh consumption and industrial supply, constituting two distinct production chains, from the varieties used, cultivation methods to final consumption (CAMARGO et al., 2006).

Due to its growing demand and the ease of adding value to the end product, the tomato plant has become one of the main vegetables produced in Brazil, with an emphasis on protected cultivation. Growing tomatoes in a protected environment has become widespread in the Southeast, especially in the state of Sao Paulo, an important vegetable producing center in Brazil. In the northeast, the largest tomato producers are the states of Pernambuco and Bahia (EMBRAPA, 2003).

This technology is emerging as an alternative for horticulturists, since it minimizes the effects of environmental variability, improving crop development, allowing production throughout the year and thus achieving higher prices on the market (SILVA et al., 2000); in addition to protecting the crop from the negative effects of wind, rain and hail, it enables considerable increases in productivity, as well as precocity, better quality and savings on inputs (GOMES et al., 1999).

Growing in a protected environment makes it possible to use substrate instead of soil, especially when it comes to soils with low agricultural suitability. The advantages of growing tomatoes in a substrate compared to soil are: higher productivity, better quality fruit, more uniformity and greater commercial value (EMBRAPA, 2011).

In addition to this growing cultivation system, fertigation has become one of the main technologies used to increase crop productivity and profitability. Although it is costly, it considerably reduces agricultural risk, without being dependent on climatic conditions and the soil's natural fertility.

The high saline concentration of the medium makes it difficult for plants to absorb water and nutrients due to the low total potential of the soil solution and the chemical competition between nutrients and salts (YEO & FLOWERS, 1989).

The most direct way to re-establish normal levels of N, P, Ca and K in the plant in saline conditions would be to increase the concentration of these nutrients in the root zone by increasing the dosage of fertilizers (CUATERO & MUNOZ, 1999). These authors state that tomato plants absorbing a greater amount of Ca and K from a saline environment will have lower Na/Ca and Na/K ratios in the leaves and, consequently, a nutritional balance closer to those grown in a non-saline environment.

Based on the premise that the nutrients K and Ca are extremely important for plant development and that their extra application can mitigate the effect of salinity on tomato crops, the aim of this work was to study some relationships between these nutrients in order to obtain satisfactory responses.

2. literature review

2.1 BOTANICAL DESCRIPTION OF TOMATOES

The tomato plant (*Lycopersicum esculentum* Mill), which belongs to the solanaceae family, has a flexible stem with many lateral branches covered in hairs and a slightly sticky substance. The flowers are small and yellow and arranged in clusters. The fruits are berry-like and can be red, yellow or pink. Their shape can vary between oval, round and flat. The size of the fruit can also be very different, ranging from a few grams (cherry tomatoes) to over a kilo in the so-called 'Oxheart' (NPLANTAS, 2012).

2.2 THE SOCIO-ECONOMIC IMPORTANCE OF TOMATOES

The production of vegetables is a branch of the agricultural economy that generates jobs from sowing to marketing, among which the tomato stands out, a fruit typical of tropical climates and of important economic value to the country (FILGUEIRA, 2008).

Currently, the largest tomato producer in Brazil is the state of Goiàs, with an estimated production of 777,453 tons, accounting for 21.4% of the total harvested in 2016, which was 3,626,421 tons (IBGE, 2016). The Northeast region contributed 22% of the national harvest in 2016, with the states of Pernambuco, Ceará and Bahia standing out, producing 110,974, 144,987 and 246,949 tons (IBGE, 2016). The state of Rio Grande do Norte still has low tomato production, with a planted area of 132 hectares in 2016, production of 3,753 tons and an average yield of 29,000 kg ha^{-1} .

The tomato is one of the ten most important vegetables on the Brazilian market as it has excellent nutritional value with high levels of vitamins A and C, as well as being rich in lycopene (CARVALHO & PAGLIUCA, 2007). It also stands out among the main crops grown in protected environments, resulting in fruit of excellent commercial quality, thus guaranteeing a higher profit, and has become one of the most significant vegetables grown in Brazil (LUZ et al., 2007).

The demand for tomatoes has been reinforced by the search for healthier foods, as they are a reservoir of various antioxidant molecules or bioactive compounds, such as ascorbic acid, vitamin E, carotenoids and phenolic compounds (GEORGE *et al.*, 2004), so tomatoes are considered to be a functional food, favoring the growth in the sale of fresh and industrialized products.

2.3 EFFECT OF SALINITY ON PLANTS

The use of saline water can cause physico-chemical changes in the soil, such as an increase in electrical conductivity, sodium content, sodium adsorption ratio and the percentage of exchangeable sodium. Dias et al. (2003) state that excess salts in the soil reduce the availability of water for plants, as well as causing problems such as the effect of specific toxic ions on plant physiology, to such an extent as to affect their yield and the quality of their production.

The response of plants under different salinity conditions is an extremely complex phenomenon, involving morphological and growth changes, as well as physiological and biochemical processes. Thus, the survival of plants in saline conditions may require adaptive processes involving the absorption, transport and distribution of ions in the various organs of the plant, as well as the compartmentalization of mineral nutrients within the cells (GARCIA et al., 2010). The changes in metabolism induced by salinity are the result of various physiological responses by the plant, including changes in growth, stomatal behavior and photosynthetic capacity.

However, under prolonged exposure to salts, when high concentrations of ions accumulate in plant tissues, they promote effects such as a reduction in stomatal conductance, CO_2 assimilation and photosynthesis, and have been attributed to limitations of both stomatal and non-stomatal origin (MUNNS & TESTER, 2008).

The tomato crop is considered to be moderately sensitive to the effects of salts with water with an electrical conductivity above 2.5 dS m^{-} 1 (AYERS & WESTCOT, 1999). Medeiros et al. (2012), evaluating the effect of soil salinity from the application of fertilizers, found that the tomato plant, cv. 'Débora Plus', had a threshold salinity of approximately 3.0 dS m^{-1} . Gomes et al. (2006), working with the IPA 6 cultivar, found a reduction in yield with salinity greater than 2.65 dS m^{-1} .

Gomes et al. (2011) studied the cv. 'Samambaia' grown in coconut fiber, using saline levels from the use of desalination rejects, and verified a threshold salinity of 3.5 dS m^{-1} .

Among the tomato yield parameters affected by salt stress, the most visible effects are a higher incidence of flower abortion, a lower number of fruits, a reduction in fruit size and a reduction in average fruit mass, resulting in losses in commercial fruit production (CAMPOS et al., 2006; MEDEIROS et al., 2012). Despite causing a reduction in fruit production, salt stress promotes an increase in the concentration of sugars, organic acids and the percentage of fruit dry mass, a fact already confirmed by Campos et al. (2006).

It can be seen that there is great variability in the tolerance of crops to salinity, which varies depending on various factors, such as fertigation management (MEDEIROS et al., 2012), the source of salts in the water (CAMPOS et al., 2006) and genetic material (AKTAS et al., 2006), 2006), and can even vary between cultivars of the same species and throughout the crop cycle (PARIDA & DAS, 2005; MUNNS & TESTER, 2008), highlighting the importance of research that enables the use of saline water in the production of these vegetables, obtaining high productivity and product quality.

2.4 PLANT NUTRITION UNDER SALT STRESS

In the Northeast, water with a high sodium concentration is common, and the excess of sodium cation in the root growth medium promotes competition for the absorption sites of other cations. This occurs as a result of ion antagonism. In^{+} on the absorption of calcium (Ca^{2+}) and potassium (K^{+}) ions, with higher ratios occurring in more sensitive plants. Na^{+}/Ca^{2+} and. Na^{+}/K^{+} (MALAVOLTA, 2006).

In vegetables of the solanaceae family, such as peppers and tomatoes, one of the main problems related to salt stress in a sodium-rich environment is the high incidence of fruit with apical rot (RUBIO et al., 2009), caused, among other factors, by the high ratio. Na/Ca^{+2+} (GRATTAN & GRIEVE, 1999), considerably reducing the production of commercial standard fruit.

Potassium is the first nutrient most accumulated by tomato plants (FAYAD et al., 2002), with the following distribution in leaves, stems, roots and fruit: 27; 18; 1 and 54%, respectively (PRADO et al., 2011). Potassium is considered a quality nutrient due to its important effects on products, affecting attributes such as color, size, acidity, resistance to transport, handling, storage, nutritional value and industrial qualities (RAIJ, 1990). According to Ho & Adams (1995), potassium is accumulated in large quantities in tomato fruit, and is required to even out ripening and increase fruit acidity, essential characteristics for good quality and taste.

In tomato growing, potassium plays a role in the synthesis of carotenoids (JOHJIMA, 1994), especially lycopene, which is responsible for the red color of the fruit, and also in the biosynthesis of sugars, organic acids, vitamin C and total soluble solids. However, when applied excessively, it can cause physiological disorders such as a drop in fruit yield and quality, a nutritional imbalance and an increase in osmotic pressure (DIAS et al., 2003).

Potassium (K) is the only monovalent cation essential for all higher plants and is the most abundant in the cytoplasm. It is not a constituent of plant organic matter, unlike N, P and S. Its importance in plant physiology stems from its high rate of mobility within the plant and its ionic activity. Its functions are very specific since it can only be partially replaced by another cation (BATAGLIA, 2005).

This nutrient (K) acts in osmotic processes, in the synthesis of proteins and in maintaining their stability, in the opening and closing of stomata, in membrane permeability and in pH control (MALAVOLTA et al., 1997). This nutrient plays an important role in tomato cultivation, considering its role in the synthesis of carotenoids, especially lycopene, which is responsible for the red color of the fruit, and also in the biosynthesis of sugars, organic acids, vitamin C and total soluble solids (JOHJIMA, 1994).

Under conditions of NaCl stress, the concentration of potassium in leaf tissue tends to decrease, negatively influencing plant growth. According to Lacerda et al. (2003), sometimes the best performance of a certain genotype under conditions of NaCl stress seems to be related to its better potassium nutrition.

Calcium is important for enzyme activation, regulating the movement of water in cells and is essential for cell division (MALAVOLTA, 2006).

According to Prado et al. (2011), calcium is the third most absorbed macronutrient by tomato plants and is distributed in the leaves, stem, roots and fruit: 70; 26; 2 and 2%, respectively. Once incorporated into the cell tissue, calcium is immobile, hence the need for a constant supply to meet fruit growth.

Apical rot is one of the most important physiological disorders in tomatoes and is considered a symptom of calcium deficiency in the fruit. It is the abiotic disorder that most influences the commercial productivity of tomatoes and is considered a key factor in avoiding losses in crop yield (LOOS et al., 2008). However, the mechanism by which external factors affect the concentration of this nutrient in the fruit and the development of apical rot, nor the type of relationship between calcium deficiency and apical rot have been unequivocally identified (SAURE, 2001).

The most likely factors for the occurrence of apical rot may be related to the high relative humidity of the air during the harvest period and the decrease in the transpiratory flow of water and nutrients to the aerial part, influencing the redistribution of $Ca+^{2}$ to the fruit, so in conditions of low relative humidity, it is possible that there is a greater accumulation of calcium in the fruit (PAIVA et al., 1998).

This disorder is caused not only by the absorption of calcium, but also by conditions that affect the distribution of calcium to the fruit, such as the low mobility of this nutrient and the high growth rate of the fruit (FONTES, 2003). According to (BLANCO et al., 2008), nutritional stress in plants under saline stress may be associated with excessive accumulation of certain ions and reduced acquisition of others, due to changes in nutrient availability, competition in the absorption process and inhibition of transport in the plant

For chili peppers, Rubio et al. (2009) found that increasing the amount of calcium in the nutrient solution led to a significant increase in the production of fruit of commercial standard, as a result of the reduction in the incidence of apical rot, although it did cause a reduction in the firmness of the fruit flesh. These studies have shown that, in general, increasing the concentration of potassium and calcium in the nutrient solution is essential for plants grown under saline stress to reach their maximum production potential.

The simultaneous application of these nutrients should be carried out with caution, as high concentrations of potassium can, through an antagonistic effect, cause a reduction in the concentration of calcium in the plant tissue, increasing the ratio K^+/Ca^{2+} (GRATTAN & GRIEVE, 1999).

Although the beneficial effects of applying potassium and calcium to tomato crops have already been observed, certain precautions need to be taken into account, such as the ratio between K ions$^+$ and Ca ions$^{2+,}$, since the ratio between them in the leaf tissue can affect the quality of the fruit (RUBIO et al., 2009).

3. MATERIAL AND METHODS

3.1 LOCATION AND CHARACTERIZATION OF THE EXPERIMENTAL AREA

The experiment was carried out in a protected environment at the Department of Environmental and Technological Sciences of the Universidade Federal Rural do Semi-Arido (UFERSA), in Mossoró, RN, located at the geographical coordinates of 5° 12' 04 S" and 37° 19' 39 W" Greenwich, with an average altitude of 18 m (Figure 1). The region's climate, according to the Koeppen classification, is of the BSwh' type (hot and dry), with very irregular rainfall, an annual average of 673.9 mm; an average temperature of 27 °C and an average relative humidity of 68.9% (CARMO FILHO & OLIVEIRA, 1995).

FIGURE 1 - Location of the experiment, Mossoró-RN, 2016

The structure of the greenhouse is made of metal cladding measuring 7.0 m x 18.0 m, with a total area of 126 m^2, a ceiling height of 4.0 m and an arched roof with transparent plastic material to protect the plants from ultraviolet rays and direct sunlight (Figure 2).

Figura 2- Vegetation house where the experiment was conducted. Mossoró-RN, 2016.

2.2 EXPERIMENTAL DESIGN AND TREATMENTS

The statistical design used was randomized blocks, in a 5 x 4 factorial scheme, with four replications, with the experimental unit represented by a pot with a capacity of 15 dm^3 , containing one plant The treatments were formed by combining five ionic ratios of potassium and calcium (F1- $K+/Ca^{2+}$ =1.5:1; F2- K $/Ca^{+2+}$ =1.25:1; F3-K $/Ca^{+2+}$ =1:1; F4-K $/Ca^{+2+}$ =1:1.25 and F5-K $/Ca^{+2+}$ =1:1.5), with four levels of irrigation water salinity (S1-0.5; S2-2.0; S3-3.5 and S4-5.0 dS m).$^{-1}$

For the lowest salinity (S1 - 0.5 dS m^{-1}), water from a deep well located on the UFERSA campus was used. To obtain the water for the other saline levels (S2, S3, S4 and S5), a mixture of NaCl, CaCl2.2H2O and MgCl2.6H2O salts was added to the lowest salinity water in water collected from a deep well located on the central campus of UFERSA, maintaining the equivalent ratio of 7:2:1 (MEDEIROS, 1992).

2.3 SETTING UP AND CONDUCTING THE EXPERIMENTS

2.3.1 CULTIVAR USED

The SUPERA F1 cultivar, used in this experiment, has the following characteristics: determined growth habit with high leafiness and high fruit set; firm periform fruits with excellent setting at high temperatures and good coloring; average size of 8 to 6 centimeters in length weighing an average of 140 grams. The cultivar has high production potential, high tolerance to cracking and can be planted all year round. Its cycle lasts an average of 110 days, making it very suitable for processing or the fresh market (AGRISTAR, 2017).

2.3.2 SOWING AND TRANSPLANTING

Seedlings were planted in expanded polyethylene trays with a capacity of 128 cells on February 12, 2015, using a commercial substrate suitable for vegetables. When the plants had about 3 to 4 definitive leaves (Figure 3A), they were transplanted into plastic containers with a capacity of 15 dm^3 , with a total of one plant in each pot.

Figura 3- Tomato seedlings and pot arrangement inside the greenhouse. Mossoró-RN, 2016.

The experiment used coconut fiber substrate (Golden Mix Granulado), made from 100% coconut fiber, with a fine texture and no basic fertilizer. This substrate was chosen because it is widely used by several researchers for studies on this cultivation system (CHARLO et al., 2011; GOMES et al., 2011).

Each pot had a drainage system at its base made up of a 2 cm layer of envelope (gravel + geotextile blanket) to facilitate the drainage of some of the salts.

The pots were arranged inside the greenhouse in 4 rows using a spacing of 1.50 m between rows and 0.50 m between pots (Figure 3B), equivalent to a population of 13,333 plants per hectare. The first two pots planted at the beginning of each row acted as a border in which the plants were managed according to treatment T1. Wooden mounds were installed inside the greenhouse, at the ends of which number 14 wires were fixed and stretched to a height of 2.00m.

3.3.3 Cultural Practices

The plants were supported by plastic ribbons held up by a wire laid parallel to the crop rows. The 'thieving' shoots were cut back weekly, so that each plant had only one main stem.

3.3.4 IRRIGATION MANAGEMENT

A drip irrigation system was adopted, using microtube-type emitters, which had previously been evaluated under normal operating conditions. The emitters were attached to the irrigation lines (polyethylene pipes).

For each type of water, an independent irrigation system was used, consisting of a reservoir (water tank with a capacity of 500 L and a Metalcorte/Eberle self-venting electric circulation pump, model EBD250076 (driven by a single-phase motor, 210 V voltage, 60 Hz frequency, normally used in washing machines), hoses (16 mm) and microtubes with an average flow rate of 2.5 L h^{-1} (Figure 4A).

Figure 4 - Irrigation system (A) used in the experiment (Boxes + Pumps);

Thermohygrometer (B). Mossoró-RN, 2016.

Throughout the experiment, the temperature and relative humidity of the air were monitored during the development of the crop, using a thermohygrometer installed inside the protected environment under a wooden shelter at a height of 1.0 m from the ground (Figure 4B).

Irrigation was managed using a timer (digital timer, model TE-2, Decorlux®) with a frequency of 6 irrigations a day, adjusting the time of each irrigation according to the crop's needs. The water consumption of the plants was not taken into account, however, at every irrigation the humidity of the substrate was raised to the maximum water retention capacity, as soon as drainage was visible in the pots.

3.3.5 FERTIGATION MANAGEMENT

The standard fertigation (F3) used in the experiment was given immediately after the seedlings were transplanted into the pots, and 10 days after transplanting, the other fertigations were given daily, following the same irrigation frequency. The amount of nutrients for 500 L of nutrient solution supplied through fertigation followed the recommendations of Castellane & Araùjo (1995) for each tomato crop in a hydroponic system. The ratio K $/Ca^{+2+}$ =1:1 corresponds to the concentration of K and Ca recommended by Castellane & Araùjo (1995), and the other nutrients were applied in similar quantities to all the treatments (Table 1).

Table 1 - Nutrient concentrations in the fertigation used in the experiment.

Nutrient concentration (mg L $)^{-1}$						
Fertigation	N	P	K	Ca	Mg	S
F1-K+/Ca^{2+} =1.5:1	184	21	372	153	43	47,5
F2-K /Ca^{+2+} =1.25:1	184	21	310	153	43	47,5
*F3-K /Ca^{+2+} =1:1	184	21	248	153	43	47,5

F4-K /Ca^{+2+} =1:1,25	184	21	248	191,2	43	47,5
F5-K /Ca^{+2+} =1:1,5	184	21	248	229,5	43	47,5

* Castellane & Araùjo (1995).

The following fertilizers were used to prepare the nutrient solutions: calcium nitrate, calcium sulphate, calcium chloride, potassium sulphate, potassium chloride, potassium nitrate, MAP and urea, as well as a micronutrient compound. For all treatments, the same doses of nutrients were applied, except for potassium and calcium, balancing the sources used.

Foliar applications of micronutrients were made at weekly intervals using a commercial product with a solid mixture of EDTA-chelated nutrients containing 0.28% Cu, 7.5% Fe, 3.5% Mn, 0.7% Zn, 0.65% B and 0.3% Mo; during the cycle, foliar fertilizations were made with calcium and boron, essential elements during the fruiting phase.

3.4 Variables evaluated

3.4.1 GROWTH CHARACTERISTICS

During the experiments, the main plant growth parameters were assessed in a non-destructive way, using the number of leaves, plant height and stem diameter. These variables were measured every 15 days after transplanting using a graduated ruler and a digital caliper.

At the end of the experiment (90 DAT), the plants were collected and transported to the Irrigation and Salinity Laboratory to be assessed for the following growth characteristics: leaf area, dry mass of stem, leaves, fruit and total.

Leaf area was determined at the end of the experiment by collecting leaf disks using a punch with an area of 6.26 mm^2 , removing 20 leaf disks per plant. The leaves and disks were packed in paper bags and placed in a forced circulation oven at 65 °C for 72 hours. They were then weighed separately on an analytical balance. To determine the leaf area, equation 1 was used.

$$AF = \frac{AD \times MSF}{MSD} \quad (1)$$

where:

AF - leaf area, cm^2/plant;

MSF - leaf dry mass, g/plant;

MSD - disk dry mass, g;

AD - disk area, cm2.

To determine dry matter, the plants were separated into stems, leaves and fruit. The samples were then packed in paper bags, labeled and placed in an oven at 65 °C with forced air circulation until they reached a constant weight. Once dry, the samples were weighed on an electronic scale, without the paper bag, and the weight was expressed in grams per plant. The total dry matter was obtained from the sum of the different parts (MST = MSF+MSC+MSFR).

3.4.2 performance characteristics

Harvests were carried out twice a week, picking the physiologically ripe fruit. At each harvest, they were counted and weighed individually and, based on this data, the yield components were determined: number of fruits, fruit production and average fruit weight per plant. The fruits were visually analyzed for the presence of lesions caused by biotic agents (such as late blight, cracks, small borer and large borer) and abiotic agents (such as apical rot and open loculus), and then classified as commercial or non-commercial.

3.4.3 QUALITATIVE CHARACTERISTICS

The commercial standard fruit was analyzed for the main quality parameters: soluble solids content (SS), vitamin C, total acidity (AT) and the ratio of total acidity to soluble solids (AT/SS).

The SS content was determined using an ATAGO PR-100 Palette digital refractometer (Attago Co. Ltd., Japan), with automatic temperature correction and a scale ranging from 0 to 32%. The samples were taken from an aliquot of the different parts of the pulp and homogenized in a blender, with the results expressed in °Brix.

The pH was determined using a Digimed DMPH-2 digital potentiometer, previously calibrated in pH 4.0 and 7.0 buffer solutions (AOAC, 1992).

The vitamin C variable was determined by neutralization titration with Tillman's solution (2,6 dichlorophenolindophenol - DFI), according to the methodology described by

Strohecker and Henning (1967). 10 grams of pulp were used, diluted in 0.5% oxyacetic acid and transferred to a 50 mL volumetric flask. Titration was carried out on a 10 mL aliquot of this solution. The results were expressed as mg of vitamin C per 100 mL of ascorbic acid pulp.

The titratable acidity was determined according to the methodology of the Adolfo Lutz Institute (1985), determined by titrating a 10 g aliquot of juice, in duplicate, to which 40 mL of distilled water was added and then titrated with a 0.02 N NaOH solution. The end point of the titration was determined using a digital potentiometer until the pH reached 8.1. The results were expressed as a percentage of malic acid.

The titratable acidity/soluble solids ratio was determined by calculating the ratio between the two variables SS and AT.

3.5 STATISTICAL ANALYSIS

The data obtained was subjected to analysis of variance, evaluating each experiment separately. The quantitative factors relating to salinity levels were statistically analyzed using polynomial regression (linear and quadratic), while the qualitative factors (potassium/calcium ratios) were analyzed using a test for comparing means, based on the Tukey test at 5% probability. The statistical analyses were carried out using the Sisvar statistical software (FERREIRA, 2008).

4. 4. Results and discussion

4.1 TOMATO GROWTH

The analysis of variance revealed a significant effect of the interaction between the S x F factors for the leaf area (LA) variables ($p<0.01$), as well as for height (HL) ($p<0.05$), with no significant effect for the leaf number (NF) and stem diameter (DC) variables. For the F factor alone, there was no significant effect for all the variables. On the other hand, the S factor alone affected all these variables at the 1% probability level (Table 2).

Table 2- Summary of the analysis of variance for stem diameter (SD), height (HHT), number of leaves (NF) and leaf area (LA) of tomato plants subjected to salinity and calcium and potassium fertigation. Mossoró-RN, 2016.

Sources of variation	GL	Mean squares			
		DC	ALT	NF	AF
Salinity (S)	3	12,84**	950,30**	120,23**	38811950,35**
Fertigation (F)	4	4,91 ns	92,22 ns	12,79 ns	1115697,65 ns
S x F	12	3,18 ns	171,66*	15,91 ns	1400025,30**
Blocks	3	3,47	194,10	3,32	2351539,53**
Waste	60	3,02	84,18	9,46	543464,23
CV (%)		12,74	17,90	16,15	23,21

ns; *; ** = not significant, significant at 5% and 1%, respectively.

As far as leaf area is concerned, there was a significant difference between fertirrigations in the plants subjected to saline solutions of 0.5 and 2.0 dS m^{-1} . For both solutions, it is possible to see the highest leaf area values in the plants submitted to F1 and the lowest value when F4 was used, which did not differ from the others. It can therefore be seen that the plants given higher doses of K performed better than those given higher doses of Ca in terms of this variable (Table 3).

Albuquerque et al. (2011), evaluating the growth and yield of fertigated pepper under different irrigation rates and potassium doses, found that an increase in the K dose had a positive effect on this variable.

For salinities of 3.5 and 5.0 dS m^{-1} , there were no differences between the fertirrigations. In general, increasing salinity has inhibited the performance of these nutrients for PA (Table 3).

Table 3 - Average values for stem diameter (SD), height (HHT), number of leaves (NF) and leaf area (LA) of tomato plants subjected to salinity and calcium and potassium fertigation. Mossoró-

RN, 2016.

Salinity	Fertigation	DC (mm)	ALT (cm	NF	AF (cm^2 plant $)^{-1}$
0.5 dS m^{-1}	F1	16,3 a	129,2 a	19 a	8548,8 a
	F2	15,7 a	122,0 a	21 a	7715.5 ab
	F3	13,9 a	127,7 a	25 a	6953,3 b
	F4	13,8 a	121,5 a	20 a	6505,7 b
	F5	13,7 a	132,0 a	25 a	7700.4 ab
2.0 dS m^{-1}	F1	14,1 a	114,0 a	21 a	6961,1 a
	F2	13,6 a	118,0 a	19 a	5992.0 ab
	F3	14,8 a	118,7 a	19 a	6126.7 ab
	F4	12,5 a	114,0 a	16 a	5683,6 b
	F5	13,1 a	124,0 a	19 a	5924.4 ab
3.5 dS m^{-1}	F1	13,1 a	107,0 a	19 a	4572,3 a
	F2	13,3 a	116,7 a	18 a	5569,2 a
	F3	14,1 a	114,2 a	17 a	5020,8 a
	F4	11,4 a	113,2 a	19 a	4576,6 a
	F5	13,7 a	112,2 a	18 a	5769,2 a
5.0 dS m^{-1}	F1	12,6 a	95,5 a	19 a	4049,2 a
	F2	13,1 a	112,0 a	15 a	4031,6 a
	F3	12,3 a	106,7 a	16 a	3810,5 a
	F4	12,8 a	101,7 a	17 a	4320,5 a
	F5	12,5 a	105,0 a	18 a	4569,8 a

Averages followed by the same letter in the columns and for each salinity level do not differ by the Tukey test ($p < 0.05$). (F1-K^+/Ca^{2+} =1.5:1; F2-K^+/Ca^{2+} =1.25:1; F3-K^+/Ca^{2+} =1.1, F4-K/Ca^{2+} =1:1.25 and F5-K/Ca^{2+} =1:5).

Figure 5A shows the effect of salinity on stem diameter according to each fertigation. For all fertirrigations, a linear decline was observed for each unit increase in salinity for this variable. The decrease was 1.033 mm for each unit increase in salinity, totaling losses of 29%. Guedes et al. (2015), studying irrigation strategies with saline water for cherry tomatoes in a protected environment, also observed reductions in this variable. In general, the K/Ca ratios were not efficient in mitigating the effect of salt.

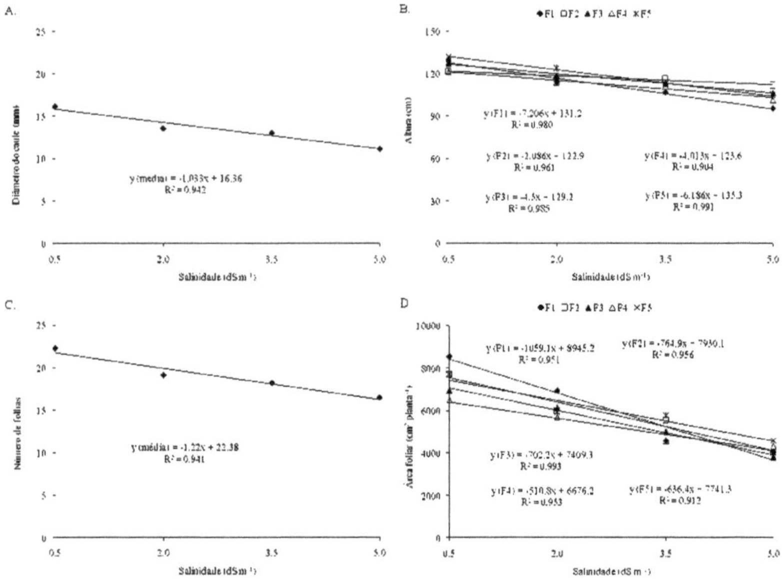

Figura 5 - Stem diameter (A), height (B), number of leaves (C) and leaf area (D) of tomato plants subjected to salinity and calcium and potassium fertigation. Mossoró-RN, 2016.

Figure 5B shows the effect of salinity on plant height according to each fertigation. It can be seen that for all the fertirrigations, the behavior was linear, with declines in plant height for each unit increase in salinity level. It can also be seen that the smallest losses occurred in the plants fertigated with F2, with a decrease of 2.086 cm, totaling a loss of 7.7%. Fertilizations F1 and F5 showed the greatest losses, reporting decreases of 7.2 and 6.2 cm per unit increase in salinity, generating total losses of 25 and 21% respectively. This behavior corroborates the results found by other authors (GUEDES et al., 2015; MEDEIROS et al., 2011), also studying the effect of salinity on tomato cultivation. In general, the application of K/Ca ratios was not efficient in mitigating the effect of salinity on this variable.

The effect of salinity on the number of leaves according to each fertigation is shown in Figure 5C. It showed that there was no positive effect of fertigation on this variable, where its behavior was linear with a decline of 1.22 leaves for each unit increase in salinity, totaling losses of 25%. In general, leaf emission was reduced as salinity levels increased in all the treatments analyzed, which shows the depreciating effect of salts on this variable.

Leaf area was affected by salinity in different ways according to the fertirrigations adopted (Figure 5D). It can be seen that salinity has a very deleterious effect on this variable, with a linear reduction for all the fertirrigations. The greatest losses were observed for F1, with a decrease of 1059 cm^2 $plant^{-1}$, causing a total loss of 56.6%. F4 and F5 showed the lowest losses, with a decrease of 510.8 and 636.4 cm^2 $plant^{-1}$, totaling 36 and 38% losses respectively.

In general, none of the K/Ca ratios were effective in mitigating the effect of salt on this variable, but it can be seen that the plants given the highest doses of Ca compared to those fertigated with the highest doses of K were better able to mitigate the salt effect.

The reduction in leaf growth represents a plant defense mechanism under conditions of water and salt stress, reducing water loss through transpiration (TAIZ & ZEIGER, 2009). However, it also represents changes in the partitioning of photoassimilates and a reduction in the area allocated to the photosynthetic process, which can result in productivity losses.

The effect of salinity on leaf area is related both to a reduction in the emission of new leaves and to the death and fall of leaves, in addition to reducing the expansion of the leaf limb (MAHMOUD & MOHAMED, 2008).

Figure 6A shows the data relating to the number of leaves throughout the experiment as a function of fertigation for each salinity level, in which

It can be seen that there was a greater difference between the fertirrigations in the plants irrigated with water with a salinity of 0.5 dS m^{-1} , where, from 25 DAT onwards, there was a greater difference, with F5 and F1 showing the greatest and least number of leaves respectively from this time onwards (Figure 6A). For the other salinities, there were no considerable variations between fertrifications, indicating that plants irrigated with saline water of 2 to 5 dS m^{-1} showed little response to variations in the K and Ca ratios (Figures 6B, 6C and 6D).

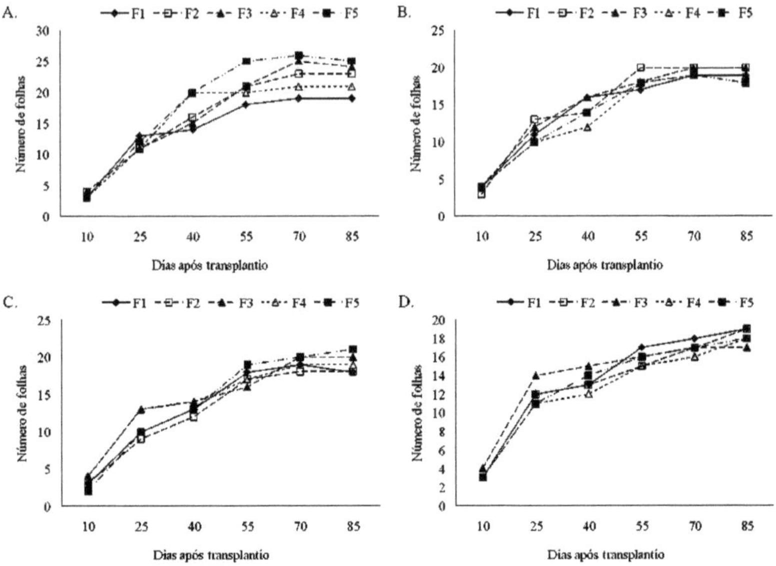

Figura 6 - Number of tomato leaves during the experiment as a function of different limestone and potassium fertigation managements and salinity levels (A - 0.5 dS m^{-1} ; B - 2.0 dS m^{-1} ; C - 3.5 dS m^{-1} ; D - 5.0 dS m^{-1}). Mossoró-RN, 2016.

Analyzing the effect of salinity on the number of leaves for each fertigation during the experiment, it can be seen that there was no difference between the salinities in the plants subjected to F1 (Figure 7A). This result may indicate that increasing the concentration of potassium in the irrigation solution reduced the effect of salinity on this variable. For the other fertrifications, there were greater variations between the salinities and, in general, the plants irrigated with the lowest salinity water 0.5 dS m^{-1} stood out from the others. It can also be seen that the difference between the 0.5 dS m^{-1} salinity and the others intensifies from 70 DAT for F2 (Figure 7B), 55 DAT for F3 (Figure 7C), and 40 DAT for F4 and F5 DAT (Figures 7D and 7E).

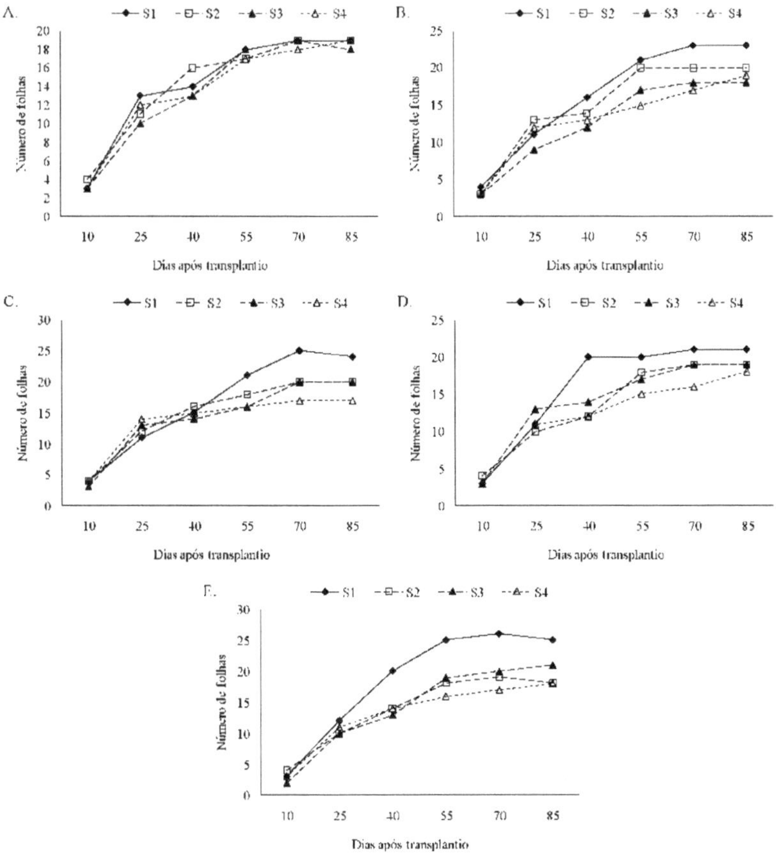

Figura 7 - Number of tomato leaves during the experiment as a function of different salinities and potassium and calcium fertilizer management (A - F1; B - F2; C - F3; D - F4; E - F5). Mossoró-RN, 2016.

Considering Figures 7D and 7E, in which the fertirrigations containing the highest concentrations of Ca, it can be seen that the increase in this nutrient in the nutrient solution increased the salt stress on this variable, although there was no greater difference at salinities above 2.0 dS m^{-1} .

Analyzing Figure 8, referring to the stem diameter throughout the experiment as a function of fertigation for each salinity level, it can be seen that there was practically no differentiation between fertigations, with the plants irrigated with water with a salinity of 0.5 dS m^{-1} standing out, where there was greater differentiation only after 85 DAT (Figure 8A). F2 stood out as having the thickest stem at the end of the cycle. For salinities above 2.0 dS m^{-1} , there was little difference in stem diameter. In general, for this variable, plants irrigated with saline water of 0.5 to 5 dS m^{-1} showed little response to variations in the K and Ca ratios (Figures 8A, 8B, 8C and 8D).

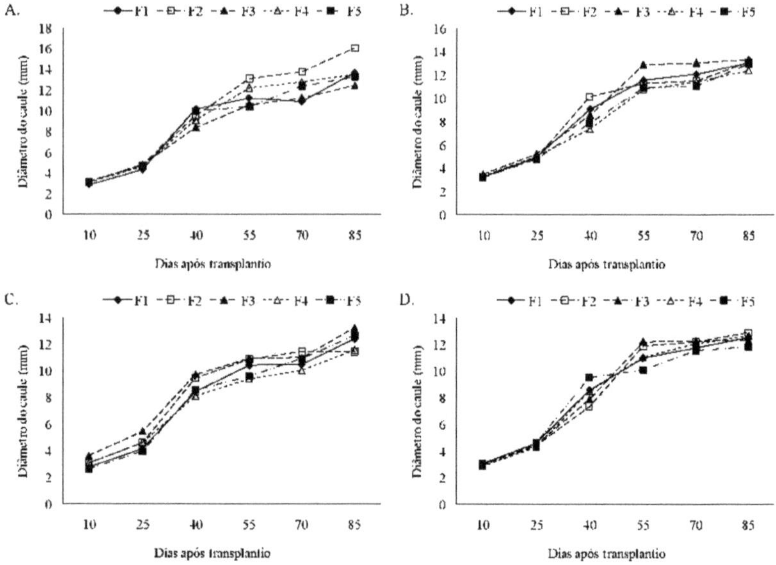

Figura 8 - Tomato stem diameter during the experiment as a function of different potassium and calcium fertigation managements and salinity levels (A - 0.5 dS m^{-1} ; B - 2.0 dS m^{-1} ; C- 3.5 dS m^{-1} ; D- 5.0 dS m^{-1}). Mossoró-RN, 2016.

Analyzing the effect of salinity on stem diameter for each fertilization during the experiment, it can be seen that there was no difference between salinities in the plants subjected to F1 (Figure 9A) and F5 (Figure 9E), these treatments showed a rapid growth rate between 25 and 40 DAT.

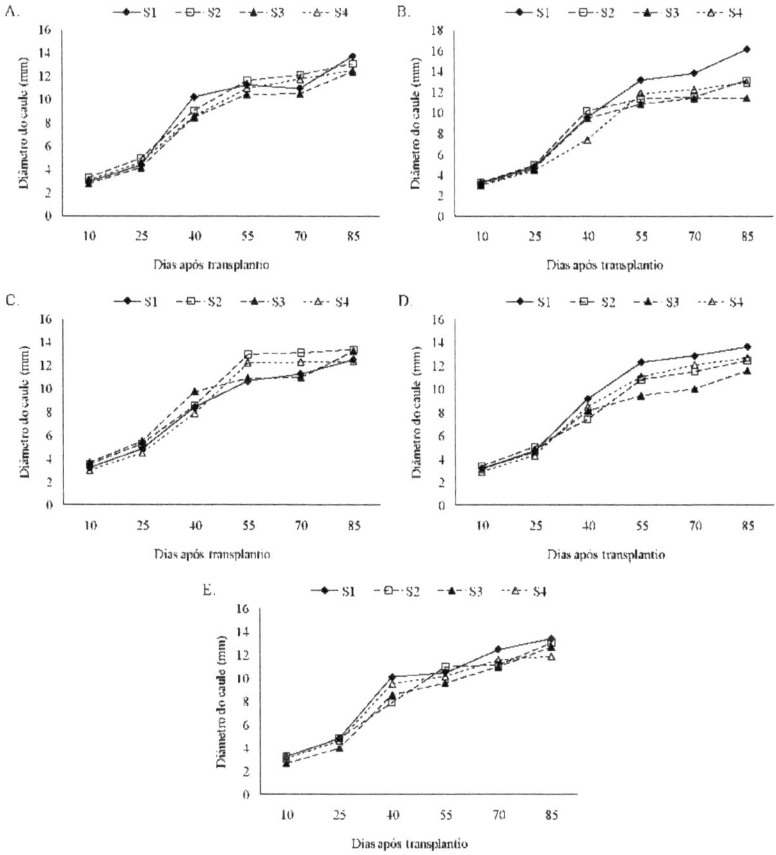

Figura 9 - Stem diameter of tomato plants during the experiment as a function of different salinities and potassium and calcium fertilizer management (A - F1; B - F2; C - F3; D - F4; E - F5). Mossoró-RN, 2016.

. These results may show that increasing the concentration of potassium and calcium in the irrigation solution reduced the effect of salinity on this variable. For the other fertrifications, there were greater variations between salinities. In general, the plants irrigated with water with a lower salinity of 0.5 dS m^{-1} were superior to the others, except for the plants fertigated with F3, which had larger diameters when subjected to a salinity of 2.0 dS m^{-1} and had accelerated growth between 40 and 55 DAT. It can also be seen that the difference between the salinity of 0.5 dS m^{-1} and the other salinities intensifies from 70 DAT onwards for F2 (Figure 9B).

The data on plant height throughout the experiment as a function of fertigation for each salinity level is shown in Figure 10. A greater difference was only observed in the plants subjected to salinity of 0.5 dS m^{-1} , in which there was a greater difference between the fertrifications from 40

DAT onwards (Figure 10A).

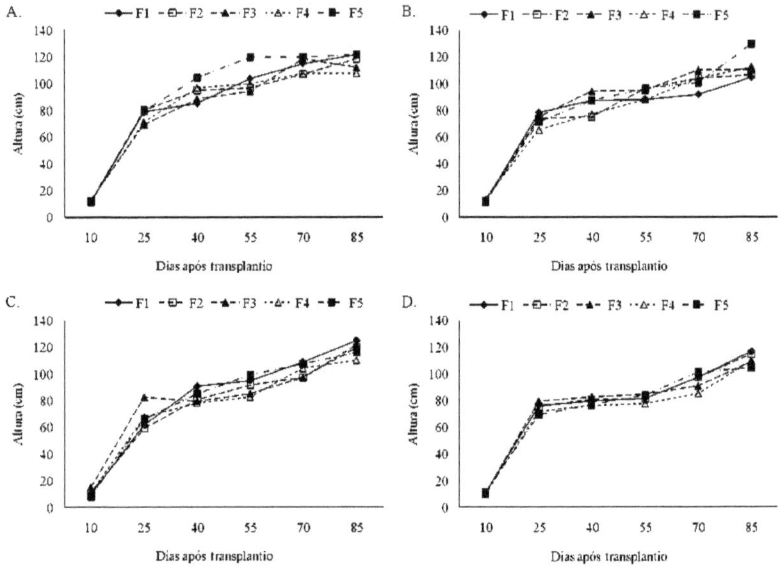

Figura 10 - Tomato plant height during the experiment as a function of different potassium and calcium fertigation managements and salinity levels (A - 0.5 dS m^{-1} ; B - 2.0 dS m^{-1} ; C - 3.5 dS m^{-1};D - 5.0 dS m^{-1}). Mossoró-RN, 2016.

In addition, the plants subjected to the highest doses of calcium stood out from the others. This could be an indication that the increase in calcium concentration favored this variable. For the other salinities, there were no significant differences between the fertrifications

Evaluating the effect of salinity on plant height during the experiment for each fertigation, it can be seen that there was a difference between salinities in all fertigations (Figure 11).

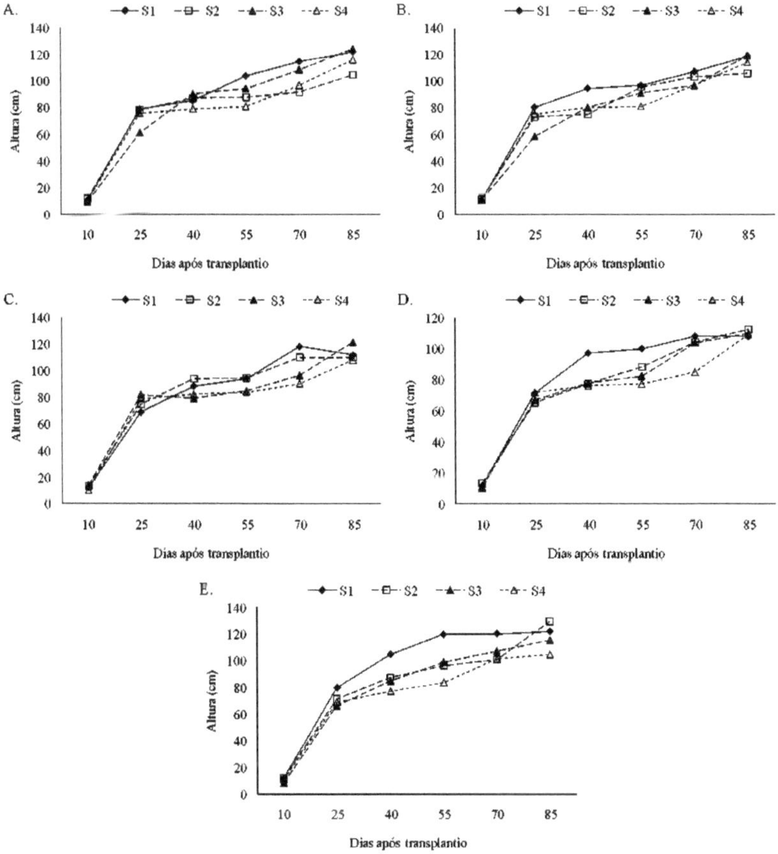

Figura 11 - Tomato plant height during the experiment as a function of different salinities and potassium and calcium fertigation managements (A - F1; B - F2; C - F3; D - F4; E - F5). Mossoró-RN, 2016.

It can also be seen that, in general, the plants submitted to solutions with the lowest salinity of 0.5 dS m^{-1} stood out with the highest values, and the difference between this salinity and the others intensified at 55; 25; 55; 25 and 25 DAT for F1 (Figure 11A); F2 (Figure 11B); F3 (Figure 11C); F4 (Figure 11D) and F5 (Figure 11E). For this variable, it can be seen that. In general, there is intense growth in plant height between 10 and 25 DAT.

4.2 DRY MATTER ACCUMULATION

From the analysis of variance, there was a significant effect of the interaction between the S x F factors for the variables leaf dry mass (LSM) and stem dry mass (SCM) ($p<0.05$), as well as for fruit dry mass (FDM) and total dry mass (TDM) ($p<0.01$). It was also observed that the MSC and MSFR variables were significantly affected at the 1% probability level by the F factor alone. The

MSF variable was affected by the F factor at the 5% probability level. For MST there was no significant effect of this factor alone, while the S factor alone affected MSF, MSC, MSFR and MST at the 1% probability level (Table 4).

Table 4 - Summary of the analysis of variance for dry mass of leaves (DMF), stem (MSC), fruit (MSFR) and total (MST) of tomato plants subjected to salinity and potassium and calcium fertigation. Mossoró-RN, 2016.

Sources of variation	GL .	Mean squares			
		MSF	MSC	MSFR	MST
Salinity (S)	3	3167,75**	1430,81**	638,61**	11839,79**
Fertigation (F)	4	130,27*	88,64**	596,31**	180,93^{ns}
S x F	12	124,74*	31,69*	489,26**	725,22**
Block	3	175,59*	130,06**	130,92**	180,93^{ns}
Waste	57	57,37	13,94	153,87	225,79
CV (%)		18,23	15,76	18,19	16,67

ns; *; ** = not significant, significant at 5% and 1%, respectively.

Table 5 shows the average values for leaf dry mass (LSM), stem dry mass (SCM), fruit dry mass (FSM) and total dry mass (TDM) as a function of fertigation management for each salinity level. For leaf dry mass, there was no difference between fertigation at salinities of 0.5; 2.0; and 5.0 dS m^{-1}, but fertigation at salinity of 3.5dSm^^{1}, in which F5 showed the highest SDF value, although it did not differ from F1, F2 and F3.

Table 5 - Average values for dry mass of leaves (MSF), stem (MSC), fruit (MSFR) and total (MST) of tomato plants subjected to salinity and calcium and potassium fertigation. Mossoró-RN, 2016.

		MSF	MSC	MSFR	MST
Salinity	Fertigation	----- g plant $^{-1}$ -----			
	F1	77,7 a	55,2 a	65,1 b	197,9 a
	F2	70,1 a	51,4 a	62,7 b	184,2 a
0.5 dS m^{-1}	F3	63,2 a	52,4 a	63,4 b	179,1 b
	F4	59,1 a	43,1 b	77,4 a	179,6 b
	F5	70,0 a	49.9 ab	68,5 b	188,4 a
	F1	51,6 a	40,5 a	76.5 ab	168.6 ab

	F2	50,6 a	42,9 a	68,1 b	159,6 b
2.0 dS m^{-1}	F3	55,7 a	41,6 a	74.2 ab	170,7 a
	F4	51,7 a	35,6 b	81,2 a	168.5 ab
	F5	59,3 a	39.7 ab	65,7 b	164.7 ab
	F1	45.1 ab	36,5 a	80,2 a	161,8 a
	F2	47.2 ab	36,1 a	70,3 b	153.6 ab
3.5 dS m^{-1}	F3	45.6 ab	38,4 a	84,6 a	168,6 a
	F4	41,6 b	33,3 a	83,9 a	158.8 ab
	F5	53,9 a	33,7 a	60,6 c	148,2 b
	F1	36,8 a	32,9 a	55,4 a	125,1 a
	F2	36,6 a	30,3 a	59,5 b	116,4 a
5.0 dS m^{-1}	F3	34,6 a	26,6 b	57,7 a	118,9 a
	F4	39,3 a	27,8 b	60,3 a	127,7 a
	F5	41,5 a	32,6 a	51,8 a	125,9 a

Averages followed by the same letter in the columns and for each salinity level do not differ by the Tukey test ($p < 0.05$). (F1-K^+/Ca^{2+}=1.5:1; F2-K^+/Ca^{2+}=1.25:1; F3-K^+/Ca^{2+}=1:1; F4- K /Ca^{+2+}=1:1.25 and F5-K /Ca^{+2+}=1:5).

For SBM, except for salinity 3.5 dS $m^{-1,}$, there was a difference between the fertrifications at the other salinities. For the 0.5 and 2.0 dS m^{-1} levels, only F4 differed from the other fertirrigations, showing the lowest values. For salinity 5.0 dS m-1 the highest values occurred in F1, F2 and F5.

With regard to MSFR, there was a significant difference between the fertrifications at all salinities. In general, it can be seen that the plants subjected to F4 had the highest MSFR for all salinities, although they did not differ from F1 and F3 at salinity 2.0 dS m^{-1} ; F1 and F4 at salinity 3.5 dS m^{-1} ; F1, F3 and F5 at salinity 5.0 dSm^{-1} . It can also be seen in Table (4) that fertigation F2 provided lower MSFR for most salinities, and was only superior to F5 at salinity 3.5 dS m^{-} 1 .

Analyzing the MST, it was found that there was a significant difference between the fertirrigations for all salinities, except for the 5.0 dS m level^{-} 1. At a salinity of 0.5 dS m^{-1} the highest values were found in F1, F2 and F5, which did not differ significantly from each other. For the 2.0 dS m^{-1} level, F3 was superior to F2, and both did not differ from the others. For the 3.5 dS m^{-1} level, F1 and F3 had the highest fertigation values, which did not differ from F2 and F4, while F5 had the lowest MST, which did not differ from F2 and F4.

Figure 12A shows the effect of salinity on leaf dry mass (LSM) according to each fertigation. All the fertirrigations showed a linear reduction in SBM, g. plant^{-1} with the greatest losses occurring in F1, F2 and F3 in the order of 8.61, 6.92 and 6.39 respectively for each unit increase in salinity, resulting in total losses of 53.7, 46.7 and 44.8% respectively. The lowest losses observed were 6.06 (F5) and 4.63 (F4), totaling losses of 39 and 35.7%. These results corroborate other studies which found a reduction in MSF when tomato plants were subjected to salt stress (GUEDES et al 2015; SILVA et al, 2013).

The effect of salinity on stem dry mass according to each fertigation is shown below in Figure 12B. For this variable, the model that fitted best was linear for all fertigation. The greatest losses occurred in F3, F1 and F2 in the following order: 5.30; 4.73 and 4.54 g plant^{-1} for each unit increase in salinity, resulting in total losses of 46.4 (F3) and 41% for F2 and F3. The lowest losses were observed in (F4) 3.21 and (F5) 3.86 g plant^{-1} , totaling 34 and 36% losses respectively.

The effect of salinity on the dry mass of the stem according to each fertigation is shown in Figure 12C. Except for F5, all the other fertigations showed quadratic behavior. In F5 there was a decrease of 3.68 g plant^{-1} for each unit increase in salinity, totaling a loss of 23.7%. F4, F3 and F1 showed the highest gains at salinities of 1.83, 2.69 and 2.54 dS m^{-1} with maximum MSFR of 84.81 and 80g plant^{-1} . Comparing these values with those obtained at the lowest salinity, there were gains of 11, 32 and 26% respectively. F2 showed the lowest value at a salinity of 2.32 dS m^{-1} with a maximum MSFR of 71g plant^{-1} . When compared to the lowest salinity, there was a gain of 15.6%.

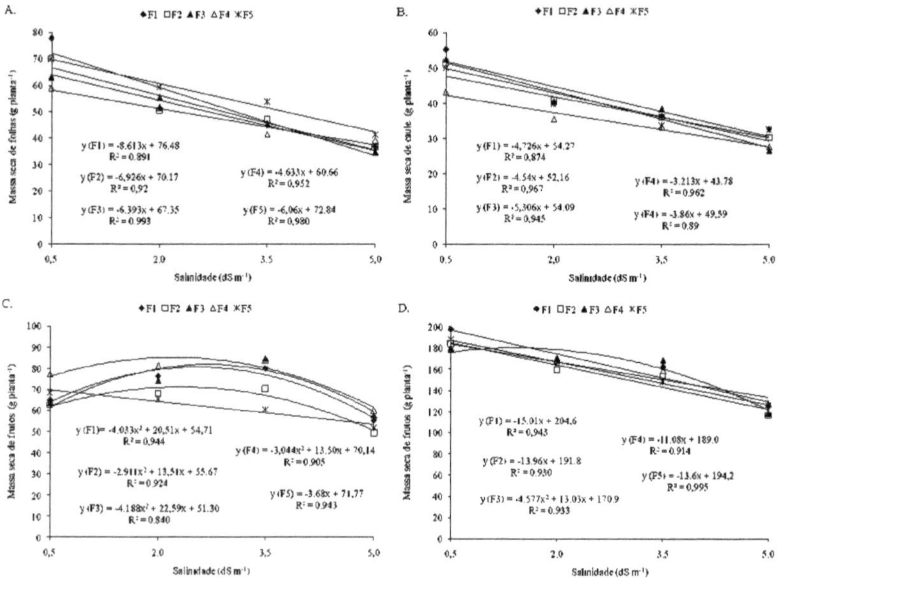

Figure 12 - Scatterplot of salinity within each fertigation for dry mass of leaves MSF (A), stems MSC (B),

fruits MSFR (C) and total MST (D) of tomato plants subjected to salinity and calcium and potassium fertigation. Mossoró-RN, 2016.

The effect of salinity on total dry mass according to each fertigation is shown in (Figure 12D). It was observed that, except for the standard fertigation (F3), which behaved in a quadratic fashion, obtaining a maximum of 180 g plant^{-1} at a salinity of 1.42 dS m^{-1} , all the others showed linear behavior. In F1 and F2, in which the plants were subjected to higher ratios of K^{+}/Ca^{2+} , there was a decrease of 15.1 and 13.96 g plant^{-1} for each unit increase in salinity respectively and they did not differ from each other, leading to losses of 34% for both in this variable. On the other hand, plants fertigated with higher doses of Ca^{2+} , showed lower losses with a decrease of 13.6 and 11.08 g plant^{-1} for each unit increase in salinity in plants subjected to F5 and F4, totaling losses of 32 and 27%.

Figure 13 shows the distribution of dry mass in the different parts of the plants. It can be seen that the plants tend to produce more dry mass in the fruit, followed by the leaves. It can also be seen that the increase in (K) in the plants subjected to the lower salinity of 0.5 dS m^{-1} showed greater production of leaf dry mass (SDF) when compared to those subjected to the higher doses of calcium. Also with regard to SBM, it was found that as salinity increased, the higher doses of (Ca) became more effective in accumulating dry mass for this variable. The behavior of stem dry mass partitioning (SCM) followed the same pattern as SCM. With regard to the distribution of the dry mass of the fruit (DMFR), it can be seen that, in general, the plants subjected to the highest doses of calcium contribute a greater percentage to the distribution of dry mass between the parts of the plant (Figure 13).

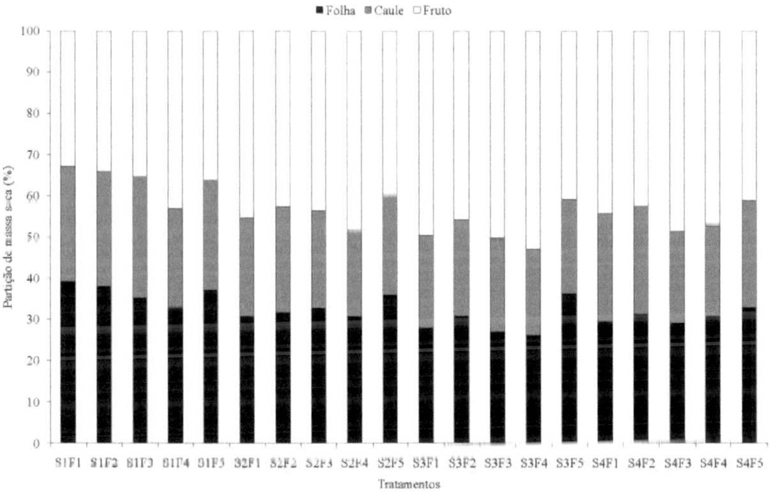

Figure 13 - Distribution of dry mass in the different parts of the plants. Mossoró-RN, 2016.

The proportions of FRM to TSS calculated from the data presented by various authors are: 66% (TEI et al., 2002), 59% (XU et al., 1995), 45% (ANDRIOLO et al., 1999) and 54 to 65% (FAYAD, 1998). Therefore, it is expected that the proportion of fruit dry matter is around 50 to 60% of the total dry matter, although these results clearly show that the proportion of dry matter produced by the different organs of the tomato plant varies according to the cultivar or hybrid, type of cultivation (soil, hydroponics, protected environment) and factors related to management. In this study, the low proportion of MSFR is an indication that the doses of K and Ca did not contribute to increasing the crop's tolerance to salinity, as the plants were unable to maintain a high proportion of fruit in relation to the proportion of other organs.

4.3 FRUIT PRODUCTION

The analysis of variance revealed a significant effect of the interaction between the S x F factors for the variables number of fruits (NFR) and yield (PROD) ($p<0.01$) as well as for average fruit mass (MMFR) ($p<0.05$). It was also found that these variables were significantly affected at the 1% probability level by the F factor alone, while the S factor alone affected NFR and MMFR at the 5% level and PROD at the 1% probability level (Table 6).

Table 6 - Summary of the analysis of variance for number of fruits (NFR), average fruit mass (MMFR) and tomato yield (PROD) subjected to salinity and calcium and potassium fertigation. Mossoró-RN, 2016.

Sources of variation	GL	Mean squares		
		NFR	MMFR	PROD
Salinity (S)	3	25,38*	145,02*	82465,82**
Fertigation (F)	4	56,35**	194,49**	77015,46**
S x F	12	36,20**	103,35*	63178,32**
Block	3	4,42^{ns}	15,88^{ns}	16902,19^{ns}
Waste	57	7,13	45,62	19869,95
CV (%)		18,80	17,16	19,96

ns; *; ** = not significant, significant at 5% and 1%, respectively.

Table 6 shows the average values for number of fruits (NFR), average fresh fruit mass (MMFR) and fruit yield PROD as a function of fertigation management for each salinity level. For the NFR, it was found that, except for the highest salinity (5.0 dS m^{-1}), there was a difference between the fertirrigations at the other salinities. When the plants were irrigated with the solution with the lowest salinity (0.5 dS m^{-1}), the highest values were obtained with fertigation F4 and F5. For salinities of 2.0 and 3.5 dS m^{-1} , fertigation F2, F3 and F4 provided the highest values (Table 6).

Table 7 - Average values for number of fruits (NFR), average mass of fruits (MMFR) and yield of tomatoes subjected to salinity and calcium and potassium fertigation. Mossoró- RN, 2016.

Salinity	Fertigation	NFR	MMFR	PROD
		unit	--- g $fruit^{-1}$ ---	--- g $plant^{-1}$ ---
	F1	20,2 b	53,7 a	765,2 b
	F2	18,5 b	43,3 b	713,3 b
0.5 dS m^{-1}	F3	20,2 b	39,2 b	720,2 b
	F4	26,3 a	35,7 b	880,1 a
	F5	25,0 a	32,9 b	760,2 b
	F1	18,5 b	50,6 a	875.4 ab
	F2	22,4 a	35,4 c	773,3 b
2.0 dS m^{-1}	F3	23,4 a	38.7 bc	856.5 ab
	F4	24,7 a	34,9 c	922,7 a
	F5	20.1 ab	40,1 b	747,8 b
	F1	17,0 b	43,2 a	911,7 a
	F2	20,7 a	34,9 b	799,2 b
3.5 dS m^{-1}	F3	22,8 a	48,4 a	945,6 a
	F4	21,0 a	45,9 a	953,3 a
	F5	17,7 b	41,9 a	688,6 b
	F1	16,7 a	42,7 a	682,1 a
	F2	16,5 a	38.8 ab	562,7 a
5.0 dS m^{-1}	F3	17,8 a	37.4 ab	656,4 a
	F4	18,5 a	37.0 ab	685,3 a
	F5	16,1 a	33,2 b	588,6 a

Averages followed by the same letter in the columns and for each salinity level do not differ by the Tukey test ($p < 0.05$). (F1-K^+ /Ca^{2+} =1.5:1; F2-K^+ /Ca^{2+} =1,25:1; F3-K^+ /Ca^{2+} =1:1; F4- K /Ca^{+2+} =1:1.25 and F5-K /Ca^{+2+} =1:5).

It can be seen that the increase in Ca concentration (F4 and F5) favored fruit set when the plants were fertigated with a lower salinity solution; however, this effect was inhibited as salinity increased. Thus, it can be seen that the effect of K and Ca on NFR depends to some extent on the salinity of the nutrient solution (Table 7), agreeing in part with the results presented by Luna et al. (2007), who reported that the action of Ca on this variable can be inhibited by salt stress.

With regard to average fruit mass, there were significant differences between the fertirrigations at all salinities. At salinities of 0.5 and 2.0 dS m-1, the highest values were obtained from plants subjected to F1. At a salinity of 3.5 dS m^{-1} , only F2 differed from the others, showing lower values. At a salinity of 5.0 dS m^{-1} it was found that the F1 and F5 fertrifications had the highest and lowest MMFR, respectively, but they did not differ from the others (Table 7).

It can be seen that the plants fertigated with the highest doses of K, compared to the standard fertigation (F3), were effective in mitigating the effects of salinity 2.0 dS m^{-1} on MMFR (Table 7).

. A similar behavior was found by Yurtseven et al. (2005), who evaluated the effect of salinity and potassium fertilization on tomato plants. On the other hand, the higher doses of Ca were detrimental to MMFR and did not mitigate the effect of salinity, as was the case with the NFR variable.

These results are similar to those obtained by Shabani et al. (2012) working with cherry tomatoes, who did not observe significant gains in MMFR as a result of increasing K and Ca concentrations in plants subjected to salt stress

For fruit production, as was observed for the NFR variable, there was no difference between the fertrifications at the highest salinity, although there was a significant response at the other salinities. The plants subjected to F4 showed a higher PROD than the other fertigations when the lowest salinity solution was used, with no significant difference between the other fertigations. For salinity 2.0 dS m^{-1} the highest yield was obtained from F4, followed by F1 and F3, which did not differ from each other. For salinity 3.5 dS m^{-} 1, the highest yields were obtained from F4, F3 and F1 (Table 7).

In general, fertigation containing higher concentrations of K did not cause a significant increase in yield when compared to standard fertigation (F3) (Table 7).

These results are in agreement with those presented by Yurtseven et al. (2005), who evaluated the effect of salinity and potassium fertilization on tomato plants and found no positive effect of increasing the dose of this nutrient on fruit production.

For cherry tomatoes, Shabani et al. (2012) found that increasing the concentration of both K and Ca did not inhibit the effect of salinity on fruit production; moreover, these authors observed that excessive concentrations of calcium caused a reduction in this variable. On the other hand, Kaya & Higgs (2003) in a study of chili peppers subjected to salinity and potassium supplementation (KNO_3) found a significant response in fruit production only in plants grown under saline stress, with an increase in production as the doses of KNO_3 increased.

Figure 14A shows the effect of salinity on the number of fruits according to each fertigation. For F1, F4 and F5, the increase in salinity caused a linear reduction in the number of fruits, with losses of 0.9 (F1), 1.806 (F4) and 1.946 (F5) fruits per plant per unit increase in salinity, resulting in a total loss of 18.1, 30.4 and 36.4% for F1, F4 and F5, respectively.

These results show that fertigation F1, F4 and F5, in which higher concentrations of K (F1) and Ca (F4 and F5) were supplied, were not effective in mitigating the effect of salinity on NFR. Furthermore, it was found that excess Ca was more harmful than excess K (Figure 14A).

Tuna et al. (2007), working with hydroponic tomato plants subjected to salinity and levels of calcium in the nutrient solution, also found that increasing the concentration of calcium did not reduce the effect of salt stress on the number of fruits.

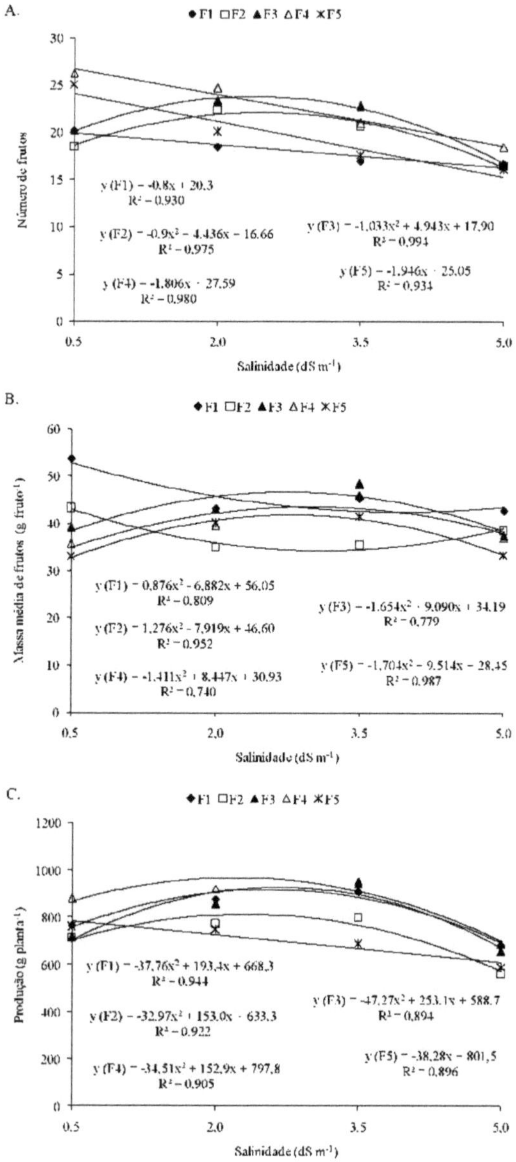

Figure 14 - Effect of salinity on the number of fruits according to each fertigation. (A) - Number of fruits; (B) - Average fruit mass; (D) - Yield. Mossoró-RN, 2016.

For F1, F4 and F5, increasing salinity caused a linear reduction in the number of fruits, with losses of 0.9 (F1), 1.806 (F4) and 1.946 (F5) fruits per plant per unit increase in salinity, resulting in a total loss of 18.1, 30.4 and 36.4% for F1, F4 and F5 respectively. For F2 and F3 there were quadratic responses with the highest values occurring at a salinity of 2.4 dS m^{-1} , with maximum NFR of 22 and 24. Comparing these values with those obtained at the lowest salinity, there were gains of approximately 18% for both.

For F2 and F3 there were quadratic responses and the highest values occurred at a salinity of 2.4 dS m^{-1} , with maximum NFRs of 22 and 24 fruits per plant, respectively. Comparing these values with those obtained at the lowest salinity, there were gains of approximately 18% for both. In this way, it can be seen that the adoption of these fertirrigations was effective in mitigating the effect of salinity on this variable.

The negative effects of salinity on the NFR of tomato plants was also reported by Silva et al. (2013) when, studying fertilizer salts and irrigation management in tomato plants, they found that by increasing the initial salinity level of the soil, the relative losses for the variable were 27.31%. Guedes et al. (2015), studying irrigation strategies with saline water for cherry tomatoes in a protected environment, also observed losses in pain production per unit increase in salinity.

The effect of salinity on the reduction in NFR may be due to changes in the osmotic potential of the substrate solution, reducing the consumption of water and nutrients by the plants, thus reducing the fruit set index (LEONARDO et al., 2008).

The effect of salinity on the average fruit mass for each fertigation is shown in Figure 14B. For all fertirrigations, it was observed that the effect behaved quadratically. The highest values occurred at salinities of 2.7dSm^{-1} and 3.0dSm^{-1} when fertigated with F3 and F4. Comparing these values with those obtained at the lowest salinity, there were gains of 21 and 25% respectively.

It can be seen that standard fertigation (F3) showed a higher MMFR, showing that the application of extra doses of potassium and calcium were not effective in mitigating the effect of salinity when compared to standard fertigation. The negative effect of salinity on MMFR has already been observed by several authors in tomato (BLANCO & FOLEGATTI, 2008; MEDEIROS et al., 2012) and other solanaceous crops, such as eggplant (OLIVEIRA et al., 2014) and pepper (LEONARDO et al., 2008).

Figure 14C shows the effect of salinity on fruit production (PROD) according to each fertigation used. There was a quadratic response except for the plants irrigated with the F5 nutrient solution, in which the increase in salinity caused a linear reduction in the crop yield variable, with a loss of 38 grams per plant for each unit increase in salinity, totaling a loss of 22%.

For fertigation F1, F2, F3 and F4 and F3 there were quadratic responses to the increase in salinity, obtaining the highest PROD at levels 2.5; 2.3; 2.7 and 2.2 dS m^{-1} , with maximum PROD of 915, 810, 927 and 967 g $plant^{-1}$, respectively. Although these fertrifications showed quadratic responses, the greatest percentage gains in PROD were obtained for F3 (31%) and F1 (21%) compared to the values obtained with the lowest salinity (Figure 14C).

It can therefore be seen that the extra doses of K^+ and Ca^{2+} were not effective in mitigating the effect of salinity on this variable. It can also be seen that the higher dose of Ca was detrimental to this variable, corroborating the results found by Tuna et al. (2007).

4.4 QUALITY

The analysis of variance showed a significant effect of the interaction between the S x F factors for the soluble solids (SS), vitamin C (Vit C), titratable acidity (AT) and soluble solids/titratable acidity (SS/AT) variables ($p < 0.01$), with no significant effect only for the hydrogen potential (pH) variable. It was also found that, with the exception of the variable (SS) ($p < 0.01$), the others did not show a significant effect due to the F factor alone, while the S factor alone affected all the variables at the 1% probability level (Table 8).

Table 8. Summary of the analysis of variance for soluble solids content (SS), Vitamin C (Vit C), titratable acidity (TA), hydrogen potential (pH) and soluble solids/titratable acidity ratio (SS/TA) in tomato fruit subjected to salinity and calcium and potassium fertigation. Mossoró-RN, 2016.

Sources of variation	GL	Mean squares				
		SS	Vit C	AT	pH	SS/AT
Salinity (S)	3	48,18**	485,89**	0,0871**	0,454**	100,07**
Fertigation (F)	4	3,49**	52,23 ns	0,0041 ns	0,123 ns	7,46 ns
S x F	12	3,75**	90,90**	0,0086**	0,110 ns	38,51**
Waste	60	0,43	23,96	0,0018	0,079	5,66
CV (%)		11,34	13,86	11,24	6,03	15,75

ns; *; ** = not significant, significant at 5% and 1%, respectively.

Table 9 shows the average values of SS, Vit. C, AT, pH and SS/AT as a function of fertigation management for each salinity level.

With regard to the soluble solids (SS) content, there were significant differences between the fertrifications at all salinities. For salinities 0.5 and 2.0 dS m^{-1} , the lowest SS values were obtained in the plants submitted to the highest calcium concentration (F5). At a salinity of 3.5 dS m^{-1} only F4 differed from the others, showing the lowest value. In the fruit obtained from plants subjected to the highest salinity (5.0 dS m^{-1}), the F1 and F5 fertirrigations had the lowest and highest values respectively, and did not differ from the other fertirrigations (Table 9).

These results corroborate those found by Eurasian et al. (2017) when they observed a decrease in SS content when CaCl2 was added to the nutrient solution, where the decrease was even more pronounced when the doses of Ca e increased. Therefore, it was found that the effect of calcium on the soluble solids content depends on the level of salinity.

Studies by other authors have found SS contents ranging from 3.8 to 5.0 °Brix (THYBO et al., 2006; PEET et al., 2004). For industrial tomato plants, Campos et al. (2006) observed SS content ranging from 4.4 to 6.9 °Brix, which increased as the salinity of the irrigation water increased. The soluble solids content in the fruit, besides being a genetic characteristic of the cultivar, is influenced by fertilization, temperature and irrigation. The average Brix values in the raw material received by industries in Brazil have been quite low (4.5 °Brix). However, there are cultivars that have greater genetic potential, showing values close to 6.0 °Brix under certain conditions (EMPRAPA, 2003).

Table 9. Average values for soluble solids (SS), vitamin C (vit C), titratable acidity (TA), hydrogenionic potential (pH) and soluble solids/titratable acidity ratio (SS/TA) of tomatoes subjected to salinity and potassium and calcium fertigation. Mossoró-RN, 2016.

Salinity	Fertigation	SS	Vit C --- g plant^{-1}	AT	pH ---	SS/AT
	F1	7,80 a	35,78 c	0,51 a	4,55 a	15,25 a
	F2	7.71 ab	46.98 ab	0,49 a	4,59 a	15,47 a
0.5 dS m^{-1}	F3	7.52 ab	49,41 a	0,51 a	4,52 a	14,46 a
	F4	7,82 a	41.13 abc	0.46 ab	4,56 a	17,18 a
	F5	6,45 b	38.78 bc	0,41 b	4,68 a	11,91 b
	F1	7,32 a	44,30 a	0.39 ab	4,69 a	18,89 a
	F2	7,62 a	37.05 ab	0,42 a	4,55 a	18,24 a
2.0 dS m^{-1}	F3	7,55 a	33,02 b	0,41 a	4,65 a	18,41 a
	F4	5,87 b	33,43 b	0,43 a	4,58 a	17,84 a
	F5	4,45 c	28,73 b	0,32 b	4,54 a	16,05 a
	F1	6,32 a	35,53 a	0,38 a	4,65 a	17,31 a
	F2	5,80 a	27,92 a	0,38 a	4,59 a	18,80 a
3.5 dS m^{-1}	F3	5,35 a	33,56 a	0.34 ab	4,76 a	15,84 a
	F4	3,55 b	34,64 a	0.33 ab	4,61 a	12,71 a

	F5	5,40 a	31,16 a	0,29 b	4,64 a	18,66 a
5.0 dS m^{-1}	F1	3,12 b	28,73 a	0,28 b	5,29 a	10,93 a
	F2	3.92 ab	35,17 a	0.31 ab	5,06 a	10,36 a
	F3	3.57 ab	33,03 a	0.31 ab	5,03 a	11,42 a
	F4	3.55 ab	29,84 a	0.32 ab	4,59 a	8,23 a
	F5	5,45 a	32,49 a	0.34 ab	4,56 a	13,99 a

Averages followed by the same letter in the columns and for each salinity level do not differ by the Tukey test ($p < 0.05$). (F1-K^+ /Ca^{2+} =1.5:1; F2-K^+ /Ca^{2+} =1.25:1; F3-K^+ /Ca^{2+} =1:1; F4- K /Ca^{+2+} =1:1.25 and F5-K /Ca^{+2+} =1:5).

The SS values found in this study are above the average for tomatoes destined for industrial processing, which is 4.5 °Brix (SILVA & GIORDANO, 2000).

In general, it can be seen that increasing the concentration of K in the fertigation solution did not lead to gains in SS content at the lower salinities, but it did lead to a reduction in SS at the higher salinities. It can also be seen that increasing the Ca concentration reduced the SS content at the lower salinities, but did not affect the SS concentration at the higher salinities.

These results indicate that, to a certain extent, the K^+ /Ca^{2+} ratios tested were not efficient in increasing the concentration of sugars in tomato fruit compared to the K /Ca^{+2+} ratio of the nutrient solution considered to be standard. These results are partly similar to those observed by Abrahao (2011), who found no effect of varying Ca^{2+} / K /Mg^{+2+} on the SS content of cherry tomato fruit.

Analyzing the vitamin C content, it was found that the plants irrigated with the highest salinities 3.5 and 5.0 dS m^{-1} showed no significant differences between the fertrifications. For the salinities 0.5 and 2.0 dS m^{-1} , there was a significant difference between the fertirrigations, where in both salinities the lowest values were found in the plants submitted to the nutrient solution with the highest calcium concentration (F5) (Table 9).

Although the variation in vitamin C content was significant at the lower salinities of 0.5 and 2.0 dS m^{-1} , with values ranging from 28.73 to 49.4, these are within the normal range for tomatoes, which according to Hobson & Davies (1971), ranges from 16 to 60 mg of ascorbic acid/100 g of fruit. Silva et al, (2013) when assessing the physico-chemical aspects of tomato fruit as a function of potassium and nitrogen doses, observed that within the estimated dose (61.5 kg of nitrogen) the vitamin C values increased as the potassium doses increased up to a maximum of 36.9 mg of ascorbic acid/100 g of fruit, this behavior can be verified in the present work where the highest vit C values can be observed where the potassium doses were higher.

For titratable acidity (TA), there was a significant effect between fertirrigations at all salinities. At salinities of 0.5, 2.0 and 3.5 dS m^{-1} , the lowest values were found in plants subjected to F5. At a salinity of 5.0 dS m^{-1} , the lowest values were found in plants subjected to F1 fertigation. It can thus be seen that the increase in K availability in relation to the standard K^{+}/Ca^{2+} ratio did not affect this variable. On the other hand, a nutrient solution with a higher Ca concentration led to a reduction in TA (Table 9).

According to Kader et al. (1977), tomato fruit for fresh consumption should have a TA above 0.32%, so, with the exception of the fruit obtained from the F5 and F1 fertirrigations, at salinities of 3.5 and 5.0 dS m^{-1} , respectively, all the other treatments had fruit that was adequate according to this quality parameter.

In general, it was found that increasing the availability of K in the nutrient solution did not improve the quality indicators of tomato fruit, even causing a reduction in SS and AT when the plants were irrigated with higher salinity water.

According to Genùncio et al. (2010), the effects of K supply on the sensory and quality parameters of tomato fruit can vary in response to different environmental conditions and the range of K doses evaluated.

There was no significant effect of fertigation on hydrogen potential (pH) at any of the salinities, with pH ranging from 4.52 to 5.29. This shows that this variable is little affected by the concentrations of K^{+} and Ca^{2+} in the nutrient solution, which is in line with the results presented by Abrahao (2011), who also found no effect of the concentrations of K^{+} and Ca^{2+} on the pH of cherry tomato fruit. The lack of effect of increasing the concentration of K^{+} in the nutrient solution on the pH of tomatoes was also observed by Genùncio et al. (2010) when evaluating the production of tomato cultivars under different N/K ratios. However, these authors (ABRAHÂO, 2011; GENÙNCIO et al., 2010) found significant differences in pH between cultivars, showing that pH is a variable inherent to genetic material.

Schwarz et al. (2013) evaluated the quality of tomato hybrids and found that the pH of the Supera F1 hybrid used in this study ranged from 4.24 to 4.44 over two years of cultivation, values close to those obtained in this study.

The fruit obtained from all the treatments can be classified as non-acidic, because according to Gould (1974), for a tomato fruit to be considered acidic its acidity must be less than 4.5. All the treatments showed fruit with a pH within or close to the range considered ideal for tomatoes, which, according to Silva and Giordano (2000) is between 3.7 and 4.5, and fruit with a pH classified as not so acidic are preferred by consumers (BORGUINI & SILVA, 2007).

The soluble solids/titratable acidity ratio (SS/AT) only showed a significant difference between the fertirrigations when the plants were subjected to the solution with the lowest salinity (0.5 dS m^{-1}), with the lowest value for the plants irrigated with the F5 fertirrigation. These results show that the increase in Ca concentration negatively affected the SS/AT ratio (Table 9).

The values found in this study are within the quality range. According to Kader et al. 1978; Mencarelli and Saltveit Junior 1988, tomato fruit is considered to be of excellent quality when it has AT values above 0.32% and SS above 3% and SS/AT ratio above 10. The SS/AT ratio has been used as an indicator of the taste of tomato fruit.

According to Jones Júnior (1999), high acid content and low sugar content result in fruit with a sour taste, while high sugar content and low acid content provide a mild taste. High values for the SS/AT ratio indicate a mild flavor due to the excellent combination of sugar and acid, while low values correlate with a sour flavor (FERREIRA et al., 2004).

The effect of salinity on SS content according to each fertigation is shown in Figure 15A. With the exception of F4, which showed a linear reduction of 0.972 °Brix in SS content for each unit increase in salinity, totaling significant losses of around 58%, the other fertirrigations showed a quadratic response to the increase in salinity. The plants fertirrigated with F1 obtained the highest values among the fertirrigations when irrigated with salinity 1.1 dS m^{-1} , with a maximum SS content of 7.82. When these values were compared to those obtained with the lowest salinity, the gain was 1.37%.

The effect of salinity on vitamin C content according to each fertigation can be seen in Figure 15B. For F2, F3 and F5 there was a reduction from the initial salinity up to the salinity levels of 3.5, 3.7 and 3.2 dS m^{-1} respectively, from which point the vitamin C content tended to increase. For fertigation F1, there was a quadratic response to the increase in salinity, obtaining a maximum value of 41.45 mg/100g of pulp in the vitamin C content when the plants were irrigated with a salinity of 2.16 dS m^{-1} . For F4, the observed behavior was a linear reduction with a decrease of 2.177 mg/100g of pulp for each unit increase in salinity.

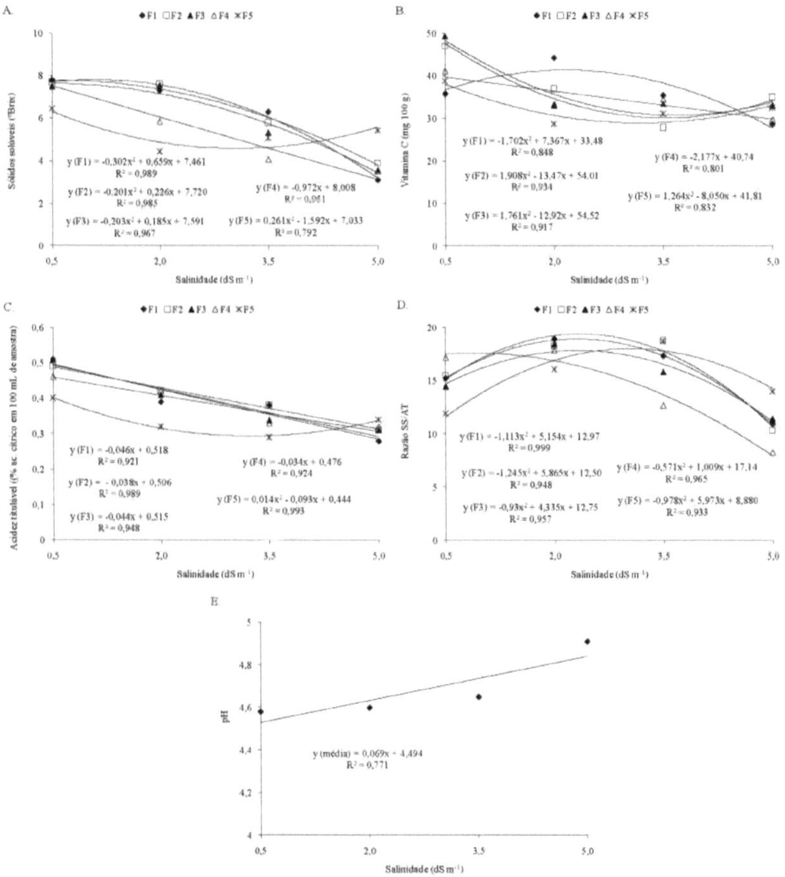

Figure 15 - Shows the effect of salinity on SS content according to each fertigation.

The effect of salinity on titratable acidity (TA) was different for each fertigation (Figure 15C). With the exception of the plants fertigated with F5, which showed a quadratic response to the increase in salinity, reaching a maximum value of 0.401 (% citric acid in 100 mL of sample) at salinity 0.5 dS m^{-1} , from which the TA value decreased until salinity 3.3 dS m^{-1} , and then tended to increase, the other salinitics showed a linear response to the increase in salinity. F1 and F3 showed the highest losses of 0.046 and 0.044 (% of citric acid in 100 mL of sample) per unit increase in salinity, resulting in a total loss of 42 and 40% respectively.

Figure 15D shows the effect of salinity on the soluble solids/titratable acidity ratio (SS/AT) according to each fertigation. It can be seen that the ratio (SS/AT) for all plants had a quadratic effect when salinity was increased. F4 had the lowest (SS/AT) value (17.6) and the lowest percentage gain (0.48%). Fertilizations F2 and F1 showed the highest values, 19.40 and 19.00, with gains of 28 and 24% respectively, when compared to the values obtained with the lowest salinity. For F5, there was a lower value of 18.4 compared to F2 and F1, but the value in terms of percentage gain, 58.6%, was much higher than both.

5. CONCLUSIONS

The salinity of the irrigation water reduced the growth variables independently of the ratios K^+/Ca^{2+}

.

Fruit dry mass was more affected by salinity when fertigation was carried out with a higher concentration of calcium.

Fertigation containing a higher concentration of calcium increased the effect of salinity on fruit number and yield.

The use of water with a salinity of 5.0 dS m-1 reduced the soluble solids content, the titratable acidity and the SS/AT ratio for most fertigation treatments.

The K /Ca^{+2+} ratios studied did not mitigate the effect of salinity on the tomato crop grown in coconut fiber.

The use of fertigation with low K /Ca^{+2+} ratios potentiates the effect of salinity on the tomato crop.

6. BIBLIOGRAPHICAL REFERENCES

ABRAHÂO, C, K:Ca:Mg ratio in the nutrient solution for growing mini tomatoes in substrate. 2011. 86 f. **Dissertation (Master's Degree in Horticulture) - Universidade Estadual Paulista "Julio de Mesquita Filho",** Botucatu, 2011.

AKTAS, H. et al. Genotypic variation in the response of pepper to salinity. **Scientiae Horticulturae**, v.110, n.3, p.260-266, 2006.

ALBUQUERQUE, F. S.; SILVA, E F. F.; FILHO, J. A. C. A.; NUNES, M. F.F. N. Growth and yield of fertigated pepper under different irrigation rates and potassium doses. **Revista Brasileira de Engenharia Agricola e Ambiental**, Campina Grande, v. 15, n. 7, p. 686-694, 2011.

ANDRIOLO, J. L.; DUARTE, T. S.; LUDKE, L.; SKREBSKY, E. C. Characterization and evaluation of substrates for growing tomato plants out of the ground. **Horticultura Brasileira**, Brasilia, v. 17, n. 3, p. 215-219, 1999.

AOAC - ASSOCIATION OF OFFICIAL ANALYTICAL CHEMISTRY... **Official Methods of Analysis of the Association of Official Analytical Chemistry.** 11. Ed. Washington: AOAC, 1992. 1115 p.

AYERS, R. S.; WESTCOT, D. W. Water quality in agriculture. Campina Grande: UFPB, 1999. 153p. FAO Studies, Irrigation and Drainage, 29

BATAGLIA, O.C. Diagnostic methods for potassium nutrition with emphasis on DRIS. In: YAMADA, T.; ROBERTS, TL. SIMPÒSIO SOBRE ADUBAÇÃO POTASSICA NA AGRICULTURA BRASILEIRA, 2, 2005, Sao Pedro, SP. **Proceedings**. Piracicaba: Brazilian Association for Potash and Phosphate Research.

BLANCO, F. F. et al. Doses of N and K in tomato plants under saline stress: I. Nutrient concentration in soil and plant. **Revista Brasileira de Engenharia Agricola e Ambiental**, v.12, n.1, p.26-33, 2008.

BLANCO, F. F. **Tolerance of tomato plants to salinity under fertigation and calibration of specific ion meters for determining nutrients in the soil solution and in the plant**. Piracicaba: ESALQ/USP, 2004. 115p. PhD Thesis

BORGUINI, R. G.; SILVA, M. V. The nutritional content of organically and conventionally grown tomatoes. **Revista Higiene Alimentar**, Itapetininga, v. 21, n. 149, p. 4146, 2007.

BRESLER, E.; MCNEAL, B. L.; CARTER, D. L. Saline and **sodic soils**: Principles, dynamics and modeling. New York: Springer, 1982. 236p.

CAMARGO, A. M. M.; CAMARGO, F. P.; ALVES, H. S.; CAMARGO FILHO, W. P.. Development of the tomato agro-industrial system. **Informações Econômicas,** Sao Paulo, v. 36, n. 6, p. 53-65. 2006.

CAMPOS, C. A. B.; FERNANDES, P. D.; GHEYI, H. R.; BLANCO, F. F.; GONÇALVES, C. B.; CAMPOS, S. A. F. Yield and fruit quality of industrial tomato under saline irrigation. **Scientia Agricola**, Piracicaba, v. 63, n. 2, p. 146-152, 2006.

CARMO FILHO, F.; OLIVEIRA, O. F. **Mossoró**: a municipality in the semi-arid Northeast, climatic characterization and floristic aspect. Mossoró: ESAM, 1995. 62p. (Mossoroense Collection, series B).

CARVALHO, J. L.; PAGLIUCA, L. G. Tomato: A market that keeps growing globally. **Revista Hortifruti,** Piracicaba, v. 6, n. 58, p. 6-14. 2007.

CASTELLANE, P. D.; ARAÙJO, J. A. C. **Soilless cultivation**: hydroponics. Jaboticabal: FUNEP, 1995. 43 p.

CHARLO, H. C. O.; OLICEIRA, S. F.; CASTOLDI, L.; VARGAS, P. F.; BRAZ, L. T .; BARBOSA, J. C.Growth analysis of sweet pepper cultivated in coconut fiber in a greenhouse. **Horticultura Brasileira**, Brasilia, v. 29, n. 3, p. 316-323, 2011.

CUARTERO, J.; MUNOZ, R.F. Tomato and salinity. **Scientia Horticulturae**, Amsterdam, v. 78, n. 1/4, p. 83-125, 1999.

INTER-UNION DEPARTMENT OF STATISTICS AND STUDIES

SOCIOECONOMICS (Dieese). **World and Brazilian tomato production**, 2010.

DIAS, N. S. et al. **Prevention, management and recovery of soils affected by salts**. Piracicaba: ESALQ, 28. Department of Rural Engineering, 2003. 118p. Didactic Series, 13.

DIAS, N. S. et al. **Prevention, management and recovery of soils affected by salts**. Piracicaba: ESALQ, 28. Department of Rural Engineering, 2003. 118p. Didactic Series, 13.

BRAZILIAN AGRICULTURAL RESEARCH Corporation EMBRAPA. **Manual of soil analysis methods**. 2.ed. Rio de Janeiro, Ministry of Agriculture and Supply, 1997. 212p.

FAYAD, J. A. Nutrient absorption, growth and production of tomato plants grown in field and greenhouse conditions. **Viçosa**: UFV, 1998. 81p. Master's dissertation.

FAYAD, J. A. et al. Nutrient uptake by tomato plants grown under field and protected environment conditions. **Horticultura Brasileira**, v.20, n.1, p.90-94, 2002.

FERREIRA, D. F. SISVAR: a program for statistical analysis and teaching. **Revista Cientifica**

Symposium, v.6, n.2, p.36-41, 2008.

FERREiRA, R. M. A.; LOpES, W. A. R.; AROuCHA, E. M. M.; MANO, N. C. S.; SOUSA, C. M. G. S. 2012. Physical and chemical characterization of tomato hybrids at different ripening stages produced in Baraûna, Rio Grande do Norte. **Revista Ceres**, Viçosa, v. 59, n. 4, p. 506-511, 2012.

FERREiRA, S. M. R.; FREiTAS, R. J. S.; LAZZARi, E. N. padrao de identidade e qualidade do tomate (*Lycopersicon esculentum*) de mesa. **Ciência Rural,** Santa Maria, v. 34, n. 1, p. 329-335. 2004.

FiLGUEiRA, F. A. R. **Manual de olericultura**. Modern agrotechnology in the production and marketing of vegetables.2.ed.rev.ampl.Viçosa:UFV, 2008. 402p..

FiLGUEiRA, F. A. R. **Novo Manual de Olericultura**: agrotecnologia moderna na produção e comercializaçao de hortaliças. 3.ed.Viçosa: UFV, 2008. 421p.

FONTES, p. C. R. tomato apical rot, leaf edge burn in lettuce and bitter fruit blight in apple: calcium deficiency? **Horticultura Brasileira**, v.21, n.2, p.145, 2003.

GARCIA, G. O. et al. Responses of bean genotypes to salinity. **Engineering in Agriculture**, v.18, n.4, p.330-338, 2010.

GENUNCIO, G. C.; SILVA, R. A. C.; SA, N. M.; ZONTA, E.; ARAUJO, A. P. Production of tomato cultivars in hydroponics and fertigation under nitrogen and potassium ratios. **Horticultura Brasileira,** Brasilia, v. 28, n. 4, p. 446-452, 2010.

GEORGE, B.; KAUR, C.; KHURDIYA, D. S.; KAPOOR, H. C. Antioxidants in tomato (*Lycopersium esculentum*) as a function of genotype. **Food Chemistry**, Washington, v. 84, n., p. 45-51, 2004.

GOMES, J. W. S.; DIAS, N. S.; OLIVEIRA, A. M.; BLANCO, F. F.; SOUSA NETO, O. N. Growth and production of cherry tomatoes in a hydroponic system with desalination waste. **Revista Ciência Agronòmica**, Fortaleza v. 42, n. 4, p. 850-856, 2011.

GOMES, L. A. A.; SILVA, E. C. DA; FAQUIN, V. **Fertilization recommendations in protected environments**. In: Minas Gerais State Soil Fertility Commission. Recommendations for the use of correctives and fertilizers in Minas Gerais: 5th Approximation. Viçosa, 1999. 360p.

GRATTAN, S. R.; GRIEVE, C. M. Salinity-mineral nutrient relations in horticultural crops. **ScientiaHorticulturae**, v.78, n.1/4, p.127-157, 1999.

GUEDES, R. A. A.; OLIVEIRA, F. A.; ALVES, R.C.; MEDEIROS, A. S.; GOMES, L. P.; COSTA, L. P. Irrigation strategies with saline water for cherry tomatoes in a protected environment. **Revista Brasileira de Engenharia Agricola e Ambiental**, Campina Grande, v. 19, n.

10, p. 913-919, 2015.

HO, L. C.; ADAMS, P. Nutrient uptake and distribution in relation to crop quality. Hydroponic and transplant production. **ActaHorticulturae**, n.396, p.33-44, 1995.

JOHJIMA, T. Carotene synthesis and calorizing in tomato of various genotypic lines. **Journal of the Japanese Society for Horticultural Science**, v.63, n.1, p.109-114, 1994.

HOBSON, G. E.; DAVIES, J. N. The tomato. In: HULME, A. C. **The biochemistry of fruits and their products.** London: Academic Press, 1971. v. 2, p. 437- 482.

ADOLFO LUTZ INSTITUTE. 1985. Analytical standards, chemical and physical methods for food analysis. Sao Paulo: **Adolfo Lutz Institute**. 533p.

BRAZILIAN INSTITUTE OF GEOGRAPHY AND STATISTICS - IBGE. **Levantamento sistemàtico da produçao agricola**, Rio de Janeiro, v.29 n.3 p.1-79. 2016. JOHJIMA, T. Carotene synthesis and calorizing in tomato of various genotypic lines. **Journal of the Japanese Society for Horticultural Science**, Kyoto, v. 63, n. 1, p. 109114, 1994.

JONES, J.B. Tomato Plant Culture: in the field, greenhouse and home garden. **New York,** CRC Press, 1999.199p.

KADER, A. A.; MORRIS, L.L.; CHEN, P. Composition and flavor quality of fresh market tomatoes as influenced by some postharvest handling procedures. **Journal of the American Society for Horticultural Science**, Madison, v. 113, n. 5, p. 742-745, 1978.

KADER, A. A.; STEVENS, M. A.; ALBRIGHT-HOLTON, M.; MORRIS, L. L.; ALGAZI, M. Effects of fruit ripeness when picked on flavor and composition of fresh market tomatoes. **Journal of the American Society for Horticultural Science**, Alexandria, v. 102, n. 6, p. 724-731, 1977.

KAYA, C.; HIGGS, D. Supplementary potassium nitrate improves salt tolerance in bell pepper plants. **Journal Plant Nutrition**, v.26, n.7, p.1367-1382, 2003.

LACERDA, C. F. et al. Solute accumulation and distribution during shoot and leaf development in two sorghum genotypes under salt stress. **Environmental and Experimental Botany**, v.49, n.2, p.107-120, 2003.

LEONARDO, M.; BROETTO, F.; VILLAS-BÔAS, R. L.; MARCHESE, J. A.; TONIN, F. B.; Regina, M. Nutritional status and yield components of pepper plants under fertigation during induction of saline stress in protected cultivation. **Bragantia**, Campinas, v. 67, n. 4, p. 883-889, 2008.

LOOS, R. A. et al. Identification and quantification of the components of production losses in a protected environment. **HorticulturaBrasileira**, v.26, n.2, p.281-286, 2008.

LUZ, J. M. Q.; SHINZATO, A. V.; SILVA, M. A. D. Comparison of conventional and organic tomato production systems in protected cultivation. **Bioscience Journal**, Uberlândia, v.23, n.2, p.7-15, 2007.

MAAS, E. V.; HOFFMAN, G. J. Crop salt tolerance - current assessment. **Journal of Irrigation and Drainage Division**, v.103, p.115-134, 1977.

MALAVOLTA, E. **Nutriçâo mineral de plantas**. Sao Paulo: Ed. Agr. Ceres. 2006. 631p.

MALAVOLTA, E.; VITTI, G. C.; OLIVEIRA, S. A. **Evaluation of plant nutritional status: principles and applications**. 2. ed. Piracicaba: POTAFOS, 1997. 319p.

MATTEDI, A. P.; SOARES, B. O.; ALMEIDA, V. S.; GRIGOLLI, J. F. J.; SILVA, L. J.; SILVA, D. J. H. In: SILVA, D. J. H.; VALE, F. X. R. **Tomate**: tecnologia de produçao. Viçosa: UFV. 2007. p.78-89.

MEDEIROS, J. F. **Quality of irrigation water used on properties assisted by "GAT" in the states of RN, PB, CE and evaluation of soil salinity**. 1992. 173f. Dissertation (Master's Degree in Agricultural Engineering) - Federal University of Paraiba, Campina Grande, PB, 1992.

MEDEIROS, J. F. Quality of irrigation water used on properties assisted by "GAT" in the states of RN, PB, CE and evaluation of soil salinity. 1992. 173f. **Dissertation** (Master's Degree in Agricultural Engineering) - Federal University of Paraiba, Campina Grande, PB, 1992.

MEDEIROS, R. F.; CAVALCANTE, L. F.; MESQUITA, F. O.; RODRIGUES, R. M.; SOUSA, G. G.; DINIZ, A. A. Initial growth of cherry tomato under irrigation with saline water in soil with bovine biofertilizers. **Revista Brasileira de Engenharia Agricola e Ambiental**, Campina Grande, v. 15, n. 5, p. 505-511, 2011.

MENCARELLI, F.; SALVEIT JÛNIOR, M. E. Ripening of mature-green tomato fruit slices. **Journal of American Society Horticultural Science**, v. 113, n. 5, p. 742-745, 1988.

MUNNS, R.; TESTER, M. Mechanisms of salinity tolerance. **Annual Review of Plant Biology**, Palo Alto, v. 59, p. 651-81, 2008.

PAIVA, E. A. S. et al. Occurrence of blossom end rot in tomato as a function of calcium dose in the nutrient solution and air relative humidity. **JournalofPlantNutrition**, v.21, n.12, p.2663-2670, 1998.

PAPADOPOULOS, I. Fertigation: **current situation and prospects for the future.** In: Folegatti, M.V. (ed.) Fertirrigation: citrus, flowers, vegetables. Guaiba: Agropecuâria, 1999. p.11-84.

PARIDA, A. K.; DAS, A. B. Salt tolerance and salinity effects on plants: a review. **Ecotoxicology and Environmental Safety**, v.60, p.324-349, 2005.

PEET, M. M.; HARLOW, C.D.; LARREA, E. S. Fruit quality and yield in five small- fruited greenhouse tomato cultivars under high fertilization regime. **Acta Horticulturae**, Tokyo, v. 659, p. 811-818, 2004.

PRADO, R. M. et al. Growth and nutrient uptake in Raisa tomato cultivars grown in a hydroponic system. **Semina**: Ciências Agrârias, v.32, n.1, p.19-30, 2011.

RAIJ, B. van. **Potassium**: need and use in modern agriculture. Piracicaba: Brazilian Association for Potash and Phosphate Research, 1990. 45 p.

RUBIO, J. S. et al. Yield, blossom-end rot incidence, and fruit quality in pepper plants under moderate salinity are affected by K+ and Ca2+ fertilization. **ScientiaHorticulturae**, v.119, n.2, p.79-97, 2009.

SAURE MC. Blossom-end rot tomato (*LycopersiconesculentumMill.*) - a calcium - or a stress-related disorder? **ScientiaHorticulturae**, v.90, n.3/4, p.193-208, 2001.

SCHWARZ, K.; RESENDE, J. T. V.; PRECZENHAK, A. P.; PAULA, J. T.; FARIA, M. V.; DIAS, D. M. Agronomic performance and physico-chemical quality of tomato hybrids in creeping cultivation. **Horticultura Brasileira,** Brasilia, v. 31, n. 3, p. 410-418, 2013.

SILVA, E. F. F; DUARTE, S. N.; COELHO, R. D. **Salinization of soils cultivated under protected environments in the State of São Paulo**. In: Folegatti, M.V. (ed.). Fertirrigation: citrus, flowers and vegetables. Guaiba: Agropecuâria, 1999. p.267-277.

SILVA, J. B. C.; GIORDANO, L. B. **Tomatoes for industrial processing**. Brasilia: EMBRAPA/Hortaliças. 2000. 168p.

SILVA, P. F.; LIMA, C. J. G. S.; BARROS, A. C.; SILVA, E. M.; DUARTE, S. N. Fertilizer salts and fertigation management in tomato production grown in a protected environment. **Revista Brasileira de Engenharia Agricola e Ambiental**, Campina Grande, v. 17, n. 11, p. 1173-1180, 2013.

STEDUTO, P. Fertigation. **Rivistadi agronomia**, Bologna, v.18, n.1, p.3-20, 1984.

STROHECKER, R. HENNING, H. M **Vitamin analysis**. Madrid: Paz Montalvo. 1967. 428 p.

TEI, F.; BENINCASA, P.; GUIDUCCI, M. Critical nitrogen concentration in processing tomato. **European Journal of Agronomy**, Amsterdam, v. 18, n. 1/2, p. 45-55, 2002.

THYBO, A. K., EDELENBOS, M., CHRISTENSEN, L. P., SORENSEN, J. N., THORUP-KRISTENSEN, K., Effect of growing systems on sensory quality and chemical composition of tomatoes. LWT. **Food Science and Technology**, Shanghai, v. 39, n. 8, p. 835-843, 2006.

TUNA, A. L.; KAYA, C.; ASHRAF, M.; ALTUNLU, H.; YOKAS, I.; YAGMUR, B. The effects of calcium sulphate on growth, membrane stability and nutrient uptake of tomato plants grown under salt stress. **Environmental and Experimental Botany,** Elmsford, v. 59, n. 2, p. 173-178, 2007.

XU, H.; GAUTHIER, L.; GOSSELIN, A. Effects of fertigation management on growth and photosynthesis of tomato plants grown in peat, rockwool and NFT. **Scientia Horticulturae**, Amsterdam, v. 63, n. 1/2, p. 11-20, 1995.

YEO, A. R.; FLOWERS, T. J. Selection for physiological characters - examples from breeding for salt tolerance. In: JONES, H. G.; FLOWERS, T. J.; JONES, M. B. (ed.). **Plants under stress**. Cambridge: Cambridge University Press, 1989. ch.12, p.217-234. YURTSEVEN, E.; KESMEZ, G. D.; ÜNLÜKARA, A. The effects of water salinity and potassium levels on yield, fruit quality and water consumption of a native central anatolian tomato species (*Lycopersicon esculantum*). **Agricultural Water Management,** Amsterdam, v. 78, n, 1-2, p. 128-135, 2005.

<https://nplantas.com/tomate-descricao-botanica/> Accessed on: 21 Jan. 2017.

http://agristar.com.br/topseed-premium/tomate-saladette-det-hib/superafl/2517// Accessed on: 21 Jan. 2017

MIX
Papier aus verantwortungsvollen Quellen
Paper from responsible sources
FSC® C105338

Printed by Books on Demand GmbH, Norderstedt / Germany